Herman Poole

The Calorific Power of Fuels

First Edition

Herman Poole

The Calorific Power of Fuels
First Edition

ISBN/EAN: 9783337025465

Printed in Europe, USA, Canada, Australia, Japan

Cover: Foto ©berggeist007 / pixelio.de

More available books at **www.hansebooks.com**

THE CALORIFIC POWER OF FUELS.

FOUNDED ON

SCHEURER-KESTNER'S
POUVOIR CALORIFIQUE DES COMBUSTIBLES.

WITH THE ADDITION OF

A VERY FULL COLLECTION OF TABLES OF HEATS OF COMBUSTION OF FUELS, SOLID, LIQUID AND GASEOUS.

TO WHICH IS ALSO APPENDED

THE REPORT OF THE COMMITTEE ON BOILER TESTS OF THE AMERICAN SOCIETY OF MECHANICAL ENGINEERS (DECEMBER, 1897); TABLES OF CONSTANTS USED.

BY

HERMAN POOLE, F.C.S.,

Member of the Society of Chemical Industry; the American Chemical Society; the American Society of Civil Engineers; the American Society of Mechanical Engineers; etc.

FIRST EDITION.

FIRST THOUSAND.

NEW YORK:
JOHN WILEY & SONS.
London: CHAPMAN & HALL, Limited.
1898.

TO MY WIFE
THIS BOOK IS AFFECTIONATELY
DEDICATED.

PREFACE.

THE books on fuels hitherto published in English, contain only a few scattered facts regarding their calorific powers, how they are obtained, and the practical use made of them. Quite frequently these books are consulted for these facts, and the information they do contain is utilized to its fullest extent. It was thought that a book especially devoted to this subject containing all the reliable data might be of interest, and in furtherance of that idea this book is published.

The work commenced as a translation of M. Scheurer-Kestner's "*Pouvoir Calorifique des Combustibles*"; but changes became necessary to adapt it to American methods and data, and it was deemed advisable to simply use the skeleton of the work and fill it in, as considered best. Even this skeleton has hardly been preserved intact, as the arrangement of much of the material has been changed, many portions omitted, many new ones supplied, and in some of the original discussions the argument has been so changed as to point nearly opposite to that advocated by M. Scheurer-Kestner.

The work embraces only that portion of calorimetric determinations having a bearing on fuel values. A concise description is given of the leading calorimeters, those most commonly used being described more fully than the others, and some examples of working and calculations are added.

Coal being the principal fuel naturally receives more space than any of the others, and most of the examples and calculations are based on results from this fuel. The other fuels are discussed briefly, some space being given to the heats of formation of the different kinds of gas, and the advantages gained by their use. A short account of theoretical flame temperatures is given, with the methods of calculating and applying the same.

The Report of the Committee on Boiler Tests, submitted to the American Society of Mechanical Engineers, in December, 1897, is published in full, as are also several of the appendices to the report. This report revises the old method of 1885, and gives the most recent methods of testing boilers and reporting the same.

A set of tables of constants used in this and allied subjects is given, and finally a collection of calorimetric and analytic data on all the kinds of fuel used. It is believed that these tables are fuller and more complete than any previously published in any language, and in collating them all available books and periodicals have been freely used. In all instances where the author was known, he has been credited with his results. Of course in such a large amount some unreliable data may have crept in, but all possible pains have been taken to exclude any such. The list of periodicals, etc., consulted will be found following the table of contents.

For help in the work, and especially the tabular matter, the author is under obligations to many. Prominent among them are Profs. R. C. Carpenter, E. E. Slosson, W. O. Atwater, and D. S. Jacobus; and Messrs. William Kent, R. S. Hale, F. L. Slocum, W. B. Day, and C. E. Emery. The Astor Library and the Libraries of the American Society of Civil Engineers and the American Society of Mechanical Engineers were freely used, and much help obtained from the librarians. Most of the cuts are from Scheurer-Kestner's book; a few were taken from Lunge and Hurter's Alkali-Maker's Handbook; some from Groves and Thorpe's work on Fuels; a few from the Reports of the American Society of Mechanical Engineers; two from Dingler's Polytechnic Journal; one from the *Scientific American Supplement;* and one from *Engineering News.*

The work has been unavoidably delayed waiting for desired data, some of which came too late to be used.

The author knows well that the book is far from perfect or complete, but it is as near so as could be made with the diverse kinds of material obtainable. Some errors, especially in the tables, may be found, which he hopes to correct in the future.

That it may be found of service and aid to others in their work on fuels is the sincere wish of the author.

HERMAN POOLE.

NEW YORK, Jan. 1, 1898.

CONTENTS.

	PAGE
PREFACE	v
CONTENTS	ix
AUTHORITIES	xiii

CHAPTER I.

FUELS 1
    Definitions. Fuels. Calorific Value. Heat of Combustion. Thermometers. Metastatic Thermometers.

CHAPTER II.

METHOD OF DETERMINING HEAT OF COMBUSTION 7
    Methods Depending on the Composition. On the Reducing Power.

CHAPTER III.

CALORIMETERS 12
    Installation. Evaluation in Water. Correction for Readings.

CHAPTER IV.

CALORIMETERS WITH CONSTANT PRESSURE 20
    Calorimeters using Air or Oxygen. Favre and Silbermann's. Alexejew's. Fischer's. Thomsen's. Carpenter's. Schwackhöfer's. W. Thompson's. Barrus's. Hartley and Junker's.

CHAPTER V.

CALORIMETERS WITH CONSTANT VOLUME 45
    Relation of Constant Volume and Constant Pressure. Andrews'. Berthelot's. Description. Working. Calculation.

CHAPTER VI.

MAHLER'S BOMB.. 57
    Description. Working. Calculation. Examples; Colza Oil, Coal, Gas, Coke. Atwater's. Kroeker's. Walther-Hempel. Witz's.

CHAPTER VII.

SOLID FUELS... 75
    Coal. Lignite. Peat. Coke. Charcoal. Wood.

CHAPTER VIII.

LIQUID FUELS.. 88
    Shale Oils. Petroleum.

CHAPTER IX.

GASEOUS FUELS.. 92
    Heat of Combustion from Analysis. Coal Gas. Gas of Gasogenes. Producer or Air Gas. Water and Mixed Gas. Natural Gas.

CHAPTER X.

CALORIFIC POWER OF COAL BURNT UNDER A STEAM-BOILER.......... 109
    Distribution of Heat. Weight of Fuel. Sampling the Fuel. Analysis of the Coal. Analysis of the Cinders. Duration of the Test. The Water Evaporated. Temperature of the Steam. Moisture of Steam. Corrections for Quality of Steam. Quality of Superheated Steam.

CHAPTER XI.

CALORIFIC POWER OF COAL BURNT UNDER A STEAM-BOILER—CONTINUED. AIR SUPPLIED AND WASTE GASES.................... 125
    Volume of Air Necessary to Combustion. Volume of Waste gases by Analysis. Gas Sampler. Analysis of Gases. Calculation of Volume from Analysis. Calculation of Volume of Air Supplied. Calculation of Weight of Waste Gases from Analysis. Volume of Waste Gases by the Anemometer. Fletcher's Anemometer. Segur's Differential Gauge. Hirn's Method. Dasymeter. Econometer. Gas Composimeter. Temperature of Waste Gases. Pneumatic Pyrometer. Carbon in Smoke.

CHAPTER XII.

CALORIFIC POWER OF COAL BURNT UNDER A STEAM-BOILER—CONTINUED. CALCULATION OF THE HEAT UNITS...................... 159
    Heat of Aqueous Vapor. Heat of Waste Gases. Heat of the Temperature. Heat of the Hygroscopic and Combustion Water. Calories of the Combustible Gases. Calories due to Soot. Distribution of Calories—Loss.
FLAME AND FLAME TEMPERATURES................................... 168
WEIGHT AND HEAT UNITS OF CARBON VAPOR...................... 173
EVAPORATIVE POWER OF FUEL 174

APPENDIX.

REPORT OF THE COMMITTEE ON THE REVISION OF THE SOCIETY CODE OF 1885, RELATIVE TO A STANDARD METHOD OF CONDUCTING STEAM-BOILER TRIALS 177
    Report of Committee. Rules for Conducting Trial. Form for Report.
TABLES... 198
FUEL TABLES.. 209
INDEX... 249

AUTHORITIES CONSULTED.

The following list contains the names of the different publications consulted to obtain data, especially for the tables. Dates are not usually given, as in many cases the entire file was used since 1868.

Alkali Reports, England.
American Engineer.
American Gas Light Journal.
American Manufacturer.
Annalen der Chemie und Physik.
Annales de Chimie et Physique.
Annales des Mines.
Australian Mining Standard.
Bayerisches Industrie und Gewerbeblätter.
Bell, Sir I. L., Chemical Phenomena of Iron-smelting.
Berichte der Deutscher Chemischer Gesellschaft.
Berthelot, Essai de Mécanique Chimique.
Berthier, Traité des Essais par la Voie sèche.
Bulletin No. 21, U. S. Dept. Agriculture.
    " University of Wyoming.
    " de la Société Industrielle de Mulhouse.
    " de la Société Chimique de Paris.
    " de l'Association des Propriétaires d'Appareils à Vapeur du Nord de la France.
Chemical News.
Colliery Guardian.
Comptes Rendus de l'Académie des Sciences.
Crookes and Röhrig, Metallurgy.
Dingler's Polytechnisches Journal.
Dufrénoy, Traité de Mineralogie.
Electrical Engineering.

Engineer.
Engineering.
Engineering and Mining Journal.
Engineering Mechanics.
Engineering News.
Groves and Thorpe, Chemical Technology, Vol. I.
Glückauf.
Ice and Refrigeration.
Iron Age.
Isherwood, B. M., Engineering Precedents.
" " Researches in Steam Engineering.
Jahrbuch der K. K. Berg-Akademie.
" für Geologie.
Johnson, W. B., Report to Congress, U. S. A., 1844.
Journal American Chemical Society.
" Canadian Mining Institute.
" Chemical Society.
" Franklin Institute.
" Society of Chemical Industry.
" Imperial Institute.
" Iron and Steel Institute.
" de l'Eclairage au Gaz.
" des Usines à Gaz.
" du Gaz et de l'Electricité.
" für Gasbeleuchtung.
" für Praktische Chemie.
" für Angewandte Chemie.
" of Gas Lighting.
Kent, William, Pocket-book.
Le Génie Civil.
Mémoires de la Société des Ingénieurs Civils.
Mineral Industry, Vol. I.
Mineral Resources, U. S. A., various volumes.
Mining Journal.
Morin and Tresca, Machines à Vapeur.
Oesterreichische Zeitschrift für Berg- und Hüttenwesen.
Peclet, Traité de la Chaleur.
Percy's Metallurgy, Fuels.
Philosophical Magazine.
Polytechnisches Centralblatt.
Progressive Age.
Proceedings: Alabama Industrial and Scientific Society.
" American Gaslight Association.

Proceedings: American Institute Mining Engineers.
" American Society of Civil Engineers.
" Institute of Mechanical Engineers.
" Institution of Civil Engineers.
Reports: British Alkali Commission.
" British Association of Gas Managers.
" Bureau of Mines, Canada.
" Department of Mines, New South Wales.
" Geological Survey, Ohio.
" Geological Survey, U. S.
" South Lancashire and Cheshire Coal Association on Boilers and Smoke Prevention, 1869.
Revista Minera.
Revue Scientifique et Industrielle.
" Universelle des Mines.
Sanitary Engineer.
Scheerer, Lehrbuch der Metallurgie.
Scheurer-Kestner, Pouvoir Calorifique des Combustibles.
Science.
Ser, Traité de Physique Industrielle.
Stahl und Eisen.
Stevens Indicator.
Thomsen, Thermo-chemie.
Transactions Newcastle Chemical Society.
Ure's Dictionary.
United States Census Bulletin, 1890.
Williams, C. W., Fuel, its Character and Economy.
Watt's Dictionary of Chemistry.
Witz, Traité théorique et pratique des moteurs à gaz.
Wurtz, Dictionnaire de Chimie.
Zeitschrift Physikalische Chemie.
" des Vereines Deutscher Ingenieure.
Zeitung Berg- und Hüttenwesen.

CALORIFIC POWER OF FUELS.

CHAPTER I.

INTRODUCTORY.

FUELS.

FUELS are those substances containing carbon, or carbon and hydrogen, which are utilized for the heat they produce upon union with oxygen. The products of this union, called combustion, are carbonic acid or carbonic acid and water. Many fuels, such as wood, peat, crude petroleum, etc., exist naturally; others, such as coke, charcoal, coal-gas, etc., are formed artificially.

The fuel *par excellence* to-day is coal. Improvements in transportation allow deliveries at points more and more remote from the mines, and the increasing demand, aided by new and improved machinery, tends to lower the cost. New locations are still being discovered, and the old ones are being worked more thoroughly and completely. A large portion of this book will be devoted to coal, other fuels being treated incidentally; and such treatment is fitting, since it is the study of coal to which the energies of physicists and engineers are still principally devoted in their researches on the calorific power of fuel.

For convenience of discussion the fuels will be divided into three general heads:

Solid fuels—coal, lignite, peat, coke, charcoal, and wood.

Liquid fuels—petroleum, shale oils, vegetable and animal oils.

Gaseous fuels—coal gas, producer gas, water gas, mixed gas, natural gas.

CALORIFIC POWER OR HEAT VALUE.

The quantity of heat generated by the combustion of a definite quantity of fuel in oxygen is called the calorific power, heat value, or heat of combustion.

The expression *calorific power* or *heat value* has a wider signification than heat of combustion. In the popular sense the former ones apply to the measure of an industrial yield as well as to the heat given off by the fuel during its complete combustion. The expression *heat of combustion*, more nearly correct from a scientific point of view, is applied, on the contrary, only to that quantity of heat generated by the substance when completely burnt; that is to say, when the carbon and hydrogen are completely changed to carbonic acid and water. The unit adopted for these quantities of heat is the Calorie and the British Thermal Unit.

The *Calorie* is the quantity of heat absorbed by the unit of weight of pure water when its temperature is increased one degree Centigrade. This unit is usually one gram or one kilogram. When it represents the atomic or molecular weight, it is called the *atomic* or *molecular calorie*, the gram being taken as the atomic unity.

The *British Thermal Unit* (B. T. U.) is the quantity of heat absorbed by one unit (usually one pound) when its temperature is increased one degree Fahrenheit. It is $\frac{1}{3.968}$ of a calorie.

A kilogram in burning generates n calories with a kilogram as unit and the Centigrade scale; a pound generates n calories with a pound as unit and the Centigrade scale (W. Kent's pound-calorie); or, whatever the weight taken, there will be generated the same number of calories, using the same unit of weight and the Centigrade scale. Hence to pass from the Centigrade scale to the Fahrenheit scale multiply by the factor 1.8, that being the ratio of the two scales.

In this work calories referred to the kilogram (kilocalories) will be used, and the calorie will be the quantity of heat necessary to raise the temperature of that amount of pure water one degree Centigrade. We will omit consideration of the variations in specific heat of water; to consider these it would be necessary to state that the initial temperature was 0° C. But, as remarked by Berthelot, "the calorie varies only to a very slight degree if we take the water at a slightly increased temperature—at 15° or 20°, for example; so that we are accustomed to regard as constant the specific heat absorbed by the water for each degree comprised in this interval of temperature, thus simplifying the calculations." We may lessen this little error by referring the calorie to a litre of water instead of a kilogram, that is, by measuring the water instead of weighing it; the weight of a litre of water diminishing from its maximum density at 4° C., while its specific heat gradually increases. The error of calculation is thus made less than the error of experiment.

HEAT OF COMBUSTION.

When the fuel contains hydrogen, its heat of combustion may be expressed in two ways. Hydrogen in burning produces water, and this water may be either condensed or in the state of vapor. The same number does not apply to both cases, since the vaporization of the water formed consumes heat, which is not given up to the calorimetric bath. We usually consider the heat of combustion, the result of the experiment made under ordinary conditions, or when the water is in the liquid state; this is the general acceptance of the term heat of combustion. Some authors, however, prefer to consider the water as vapor.

It is easy, however, to change from one system to the

other. The heat of combustion of one kilogram of hydrogen being 34500 calories,* and the water formed being liquid at 0° C., a portion of the 34500 calories is used to vaporize the water in the case where it is gaseous or considered as such.

Experiment has shown that the heat of vaporization of water is expressed by the formula of Regnault,

$$606.5 + 0.305t, \quad \text{or}$$

$$1091.7 + 0.305(t - 32°) \text{ for Fahrenheit degrees,}$$

in which t represents the temperature of the water in the state of vapor. Now one kilogram of hydrogen produces nine kilograms of water. To keep these nine kilograms of water in vapor, at 100° C. for example, there will be needed, by the above formula, 637 calories per kilogram of water, or nine times as much per kilogram of hydrogen, which is 5733 calories. These 5733 calories reduce to 5453 when the water is considered as being at 0° C. instead of at 100° C. Deducting 5453 calories from 34500 calories representing the heat of combustion of hydrogen, the water formed being condensed, we obtain 29047, which number represents the heat of combustion of hydrogen, the water being in the state of vapor at 0°. We will call it, in round numbers, 29100† calories, as is done by several writers.

THERMOMETERS.

Before taking up the study of calorimeters, we must consider the calorimetric thermometer, which is a most important part of the apparatus employed. The reading of the thermometer and the corrections are quite delicate and also very important, the calculation of the heat of combustion depending principally on their accuracy.

In this work calorimetric questions relating to fuel only will be considered; hence a description of ordinary ther-

---

* 62100 B. T. U.   † 52380 B. T. U.

INTRODUCTORY.

mometers and their manufacture will not be needed. They are usually bought all finished, and should be obtained only from reliable dealers.

Favre and Silbermann employed a thermometer of their own design, divided into $\frac{1}{10}$ degrees and graduated from 32° to 0° C. Each degree occupied about 0.3 inch. By means of a cathetometer they read to $\frac{1}{100}$ of a degree. Their calorimetric bath of 2 litres capacity was subjected to at least 8° elevation in temperature, and the quantity of substance necessary to use at times exceeded 2 grams. To lessen this amount of rise in temperature and also the time of combustion, they used longer thermometers, with scales reading to $\frac{1}{500}$° or even to $\frac{1}{1000}$°. Scheurer-Kestner used a thermometer divided to $\frac{1}{50}$° with his Favre and Silbermann calorimeter. Since then they have been used generally. Such thermometers are difficult to work with, and require care in manipulation, and often a series of thermometers or at least two with scales in sequence are employed. If the initial temperature of a calorimetric bath is found a little above the highest graduation on the first thermometer, and if the rise in temperature of the bath amounts to two degrees, we must

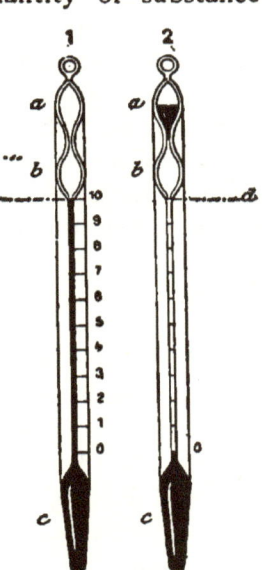

FIG. 1.—METASTATIC THERMOMETER.

substitute the second one having for its lowest degree the highest of the first. Besides the trouble of substitution, it necessitates a correction for agreement of the degrees common to the two instruments. To obviate this difficulty the "metastatic" thermometer was invented by Walferdin and described in the *Comptes Rendus de l'Académie des Sciences*, 1840, p. 292, and 1842, p. 63.

The metastatic thermometer is a differential thermometer with a variable scale. At will, a certain quantity of mercury flows into the bulb. By this means we raise or lower the degrees for which it may be used. Suppose an ordinary thermometer graduated from 0° to 10°, and left open at the top at the 10th degree. If we wish to use it between 12° and 14°, heat it to 14°, and a portion of mercury corresponding to 4° escapes. Now, instead of showing a difference of 10° between 0° and 10°, it will show this difference between 4° and 8°, the original 0° having descended to — 4°. It will be similar for temperatures of 10°, 20°, or 30°, as desired. By closing the thermometer at the top instead of leaving it open, and blowing a bulb in the upper portion as overflow, the conditions will remain the same. The thermometer has now become metastatic. These thermometers are made by Baudin of Paris, from whom full directions for use and corrections can be obtained.

With all thermometers it is essential that the glass of the bulb should be rather thin, or the thermometer will be "too slow." The slightest difference in temperature must be shown immediately by a movement of the mercurial column. To test for sensibility, read the height of the column and then place the hand on the bulb. If sufficiently sensitive the mercury will descend quickly from the expansion of the glass and afterwards rise. In thermometers divided to $\frac{1}{100}$° this movement should be immediate, and over several hundredths.

In ordinary calorimetric experiments the correction due to length of the mercury column flowing out of the bulb may be neglected for several reasons; the experiments should be made in a room where the temperature is nearly the same as that of the calorimetric bath, such correction would be of very little consequence for a slight change of temperature, and the experimenter should plunge the thermometer into the bath as deep as is necessary to take the reading at the level of the eye.

CHAPTER II.

METHODS OF DETERMINING HEAT OF COMBUSTION.

THERE are two methods for determining tne heat of combustion of substances—one by calculation based on the chemical composition, and the other by actual combustion in a calorimeter. The first method may be considered under two heads: that in which the units are calculated directly from the composition, and that in which they are calculated from the quantity of oxygen consumed during combustion in a crucible.

CALCULATION FROM CHEMICAL COMPOSITION.

Dulong stated that the heat generated by a fuel during combustion was equal to the sum of the possible heats generated by its component elements, less that portion of the hydrogen which might form water with the oxygen of the fuel.

His formula was

$$x = 8080C + 34500\left(H - \frac{O}{8}\right),$$

or expressed in B. T. U.'s,

$$x = 14500C + 62100\left(H - \frac{O}{8}\right).$$

in which
    $x =$ the heat of combustion sought;
    $8080 =$ the heat of combustion of carbon in calories;
    $14500 =$ " " " " " " " B. T. U.;
    $34500 =$ " " " " " " hydrogen in calories;
    $62100 =$ " " " " " " " " B. T. U.;

$H - \dfrac{O}{8} =$ the quantity of hydrogen less that supposed to form water with the oxygen.

Other authors and experimenters have tried to interpret their results by a general formula with varying success. Many of them by working on a certain number of coals from a certain location work out a formula which applies to that set of coals, but not as well to another set. A few of them will be given. They all resemble Dulong's and are usually only modifications of his original one.

The Verein Deutscher Ingenieure adopted the following:

$$x = 8100C + 29000\left(H - \dfrac{O}{8}\right) + 2500S - 600E,$$

in which allowance is made for the heat of combustion of sulphur and the heat of the hygroscopic water. All the coefficients are round numbers and that for hydrogen, 29000, is the one in which the water is supposed to be as aqueous vapor, all the water being considered as passing off in that state. None of the other formulæ uses this coefficient. It gives rather low results. The question as to the advisability of reckoning the heat due to sulphur is a debatable one. In no case does it amount to more than a very small per cent and can have but little effect on the total.

Balling gives as formula

$$x = 8080C + 34462\left(H - \dfrac{O}{8}\right) - 652(E + 9H)$$

to represent the actual occurrences in a steam-boiler fire working under a pressure of steam corresponding to 300° F.

Schwackhoefer made the following modification to allow for the correction due to hygroscopic water:

$$x = 8080C + 34500\left(H - \dfrac{O}{8}\right) - 637E.$$

Mahler formulated one based on the results of calorimetric determination of the heat of combustion of 44 different kinds of fuel. It is

$$x = \frac{8140C + 34500H - 3000(O+N)}{100};$$

or simplified,

$$x = 111.4C + 375H - 3000;$$

or in B. T. U.'s,

$$x = 200.5C + 675H - 5400.$$

With the coals he examined he found a very close agreement between the results calculated by this formula and those observed. A similar but not equally close concordance was found using the Dulong formula. With wood and lignites the difference amounted to 2 per cent. His formula applies also to other substances whose constituents are accurately known. Cellulose, the heat of combustion of which according to Berthelot is 4200 calories, by Mahler's formula is 4264.

In summing up he says: "From a scientific point of view, in the present state of our knowledge on the subject, we cannot give a general formula depending strictly on the chemical composition which will give the calorific power of combustibles, substances so complex and varied."

Lord and Haas in a paper read before the American Institute of Mining Engineers, Feb. 1897, state that in a series of forty Pennsylvania and Ohio coals they found differences varying from $+2.0$ to -1.8 per cent between the calculated and the observed results, and an average difference of -0.12 per cent.

In 1896 Bunte published some analyses and calorimetric tests of gas-cokes, showing a difference of from $+0.04$ to -1.2 per cent.

Three elements enter into these cases, the analysis, the calculation, and the combustion; all may be erroneous. As the matter stands now the weight of error seems to be on the side of the analysis, as our methods of analysis, especially in water determinations, are not entirely satisfactory; yet it must be confessed that some of the most recent analyses give a basis from which very close agreement can be calculated. With such fuels as coke, charcoal, or anthracite, having but little volatile matter, the results agree quite well, but with the bituminous coals, asphalts, mineral oils, etc., which are so very complex, the differences are greater.* In these the actual proximate chemical constitution seems to make a difference. It may be safely stated, however, that for ordinary industrial uses, in absence of the possibility of a calorimetric test, and with coals having under 20 per cent of volatile matter, a fairly accurate approximation may be arrived at by calculation.

The great inducement that formerly existed in favor of calculated results exists no longer. I refer to the difficulty of making a calorimetric test. These can be made now by means of the modern apparatus, so simple and almost self-regulating that the time consumed is but a small fraction of that needed for an analysis, and the labor and care, hardly anything in comparison.

If possible, by all means have a calorimetric test. If not possible, use the best analysis available.

CALCULATION FROM QUANTITY OF OXYGEN USED.

This is the litharge reduction test. It depends on Welter's formula, which is based on the hypothesis that the heat of combustion is proportional to the quantity of oxygen consumed:

$$N = mP,$$

---

* Mahler's limit for Dulong's formula is $O + N > 15$.

in which N is the heat of combustion sought, m is the coefficient previously determined, and P is the weight of oxygen necessary for the combustion of one kilogram of the substance.

Giving P the value resulting from the use of the equivalents—16 for oxygen to burn 6 of carbon, and 8 for oxygen to burn 1 of hydrogen—we have

$$P = \frac{16}{6}C + 8H = 8\left(\frac{C}{3} + H\right);$$

and the general formula becomes

$$N = 8m\left(\frac{C}{3} + H\right) = 26880\left(\frac{C}{3} + H\right).$$

To use this method the combustible is mixed with an excess of litharge and heated in a crucible. The button of lead formed shows the amount of oxygen consumed, and from this is deduced the heat by means of the formula. The heat should be increased very slowly. Mitchell substituted white lead for litharge and claimed to obtain uniform results.

This formula was recommended by Berthier, and has been used since by a few others. It is faulty, as was shown by some of Berthier's own determinations in which contradictory results were obtained. Dr. Ure showed that no uniform results could be obtained using the same materials. Scheurer-Kestner in 1892 showed that the formula not only gave erroneous results, but actually reversed the relation of combustibles. In one case cited the heats actually obtained by a calorimeter were 8813 and 8750, while by the litharge test they were 7547 and 7977. The results were not only low, but reversed the ratio.

This method is allowable only in cases where the crudest approximations are desired and where no analyses or calorimetric tests can possibly be made.

CHAPTER III.

CALORIMETRY.

CALORIMETERS for rapid combustion are invariably composed of a combustion-chamber and a calorimetric bath, usually a cylinder, surrounding it and containing a known quantity of water, the elevation in temperature of which is measured. The combustion is made in oxygen, pure or diluted.

Combustion-chambers are either under a constant pressure, as in the calorimeters of Rumford, Favre and Silbermann, etc.; or with a constant volume, as in the calorimeters of Andrews, Berthelot, etc. With solids the difference of results obtained under constant volume and constant pressure is so small that we shall not consider it. With gases, however, it is different, and we will state under which conditions the results have been obtained.

The first calorimetric experiments date from Lavoisier and Laplace. In 1814 Count Rumford replaced the ice calorimeter of Lavoisier by an apparatus in which the heat developed during the combustion was absorbed by water. It was some time after, 1858, that Favre and Silbermann discovered the causes of the great errors of their predecessors, and published methods for correcting some while avoiding others. We owe to them, above all, the observation that, even when supplied with pure oxygen, combustion may be only partial, on account of the formation of combustible gases. They determined that this occurs generally, and gave a method of estimating the unburnt gases, so as to make allowances in the calculation.

Carbon, which, before their time, had given only 7624 calories to Laplace, 7386 to Clément-Desormes, 7915 to Despretz, 7295 to Dulong, and 7678 to Andrews, yielded to F. & S. 8081 after correction for carbonic oxide in the waste gases. This number has since been increased to 8140 by the latest determinations of Berthelot. Berthelot and Vielle have shown that by using oxygen under pressure complete combustion can be attained.

INSTALLATION OF APPARATUS.

The apparatus should be placed in a room free from sudden changes in temperature and consequently protected from direct sunlight. If it is not entirely protected from solar radiation, the apparatus may be set up on the north side and shaded from the direct midday sun by a screen.

The calorimeter cylinder with its accessories, as well as the distilled water used, should remain in the room long enough to acquire its proper temperature. The cylinder should be protected as much as possible from radiation by envelopes which vary according to circumstances. Favre and Silbermann used a cylinder with a double wall. The external one was filled with water, and between this one and the cylinder proper swan's down was packed. The upper part of the cylinder also had a layer of thick paper covered with down on the under side.

Berthelot states that the down is more troublesome than useful, and that it may be omitted with advantage. The space between the cylinder and its envelope forms a layer of air which is an excellent non-conductor. In modern instruments the down is replaced by a thick layer of felt. Berthelot even omits this covering, stating that the great cause of loss of heat was not from radiation, but due to evaporation produced by the agitation of the water in contact with the air. He surrounds his cylinder with a layer of air inside of the envelope of water, and outside of all a layer of felt 0.8 inch thick. By this means external influence is much reduced.

EVALUATION OF THE CALORIMETER IN WATER.

Before using a calorimeter its equivalent in water must be determined; that is, we must calculate to what quantity of water it corresponds in terms of specific heat. This is to be added to the weight of water employed and includes the combustion-chamber, cylinder, and the immersed pieces, thermometer, supports, etc.

Below is given an example showing the calculation of the value in water of a Favre and Silbermann's calorimeter:

Copper, 1145.651 grams at 0.09516 specific heat.......... = 109.008 grams.
Platinum, 22.810 " " 0.0324 " " .... = 0.706 "

Value in water of the chamber and accessories = 109.714 "
Thermometer, weight of glass immersed, 12 grams at 0.198 = 2.400 "
Mercury, 63 " " 0.332 = 2.070 "

Total equivalent of water.................... = 114.184 "

which added to the 2 kilograms of water in the bath makes a total of 2114.184 grams of water.

The calorimetric weight for the Berthelot bomb at the College of France in 1888 was 398.7 grams for bomb and accessories.

The water value of the calorimeter used by Lord and Haas at the Ohio State University, Columbus, O., was determined as 465 grams. Mahler's apparatus had a water equivalent of 481 grams. Still, it is better to determine this equivalent by actual experiment, as we are not sure of the specific heat of the metal of the bomb, which might, however, be determined by a sample taken from the original block of which it was made.

Several methods may be employed for this.

When we use the calorimetric bomb, we burn in the obus, using 2000 grams of water, a known quantity of a substance of fixed composition, and of which the heat of combustion is known, as sugar, or naphthalin. We then use less water and burn a smaller quantity of the substance. If 1 gram of substance was taken the first time, we may take 0.8 gram with 1800 grams of water the second time. We then have two equations, from which we eliminate the heat of combustion of the substance and deduce thence the value in water of the cylinder, etc.

This method, suggested by Berthelot, may be replaced by the following, to which he gives the preference:

Pour into the calorimeter a certain quantity of warm water, at 60° C. for instance. This water is previously contained in a bottle, and the temperature is measured by a thermometer placed inside. As control, operate first without the bomb in the cylinder and afterwards with it in place.

One test of this kind gave Berthelot a value of 354 calories for the bomb. The value deduced by calculation from specific heat was 355.4. Below is the detailed calculation giving the separate parts of the bomb.

Names of the Different Parts.	Soft Steel.		Platinum.		Brass.	
	Weight in Grams.	Value in Water.	Weight in Grams.	Value in Water.	Weight in Grams.	Value in Water.
Crucible....................	1709.7	187.61	728.8	23.63		
Cover.......................	221.2	24.28	528.8	17.15		
Stop-cock...................	11.7	1.28	....	....	20.0	1.86
Cone-screw and socket of fire-carrier...........					3.97	0.37
Movable accessories serving for suspension and kindling.................			33.0	1.07		
Screw of bomb.............	802.7	88.08				
Movable foot of bomb.....	....	....	....	....	108.9	10.13
Totals...................	2745.3	301.24	1290.6	41.85	132.9	12.36

RECAPITULATION.

Metals Used.	Weight in Grams.	Calculated Value in Water.
Steel..	2745.3	301.24
Platinum..	1290.6	41.85
Brass (calorimeter and agitator omitted)........	132.9	12.36
Weight of bomb.................................	4168.8	355.45
Value in water by direct test....................	....	354.7

CORRECTIONS FOR THE READINGS.

The corrections to be applied to thermometric readings, besides those due to the thermometer itself, are of various kinds, and naturally vary with the kind of calorimeter used. Some, however, are common to all.

The correction relative to heating and cooling concerns all calorimeters. Favre and Silbermann made this correction with a coefficient previously determined, once for all, by a series of experiments. For example, the coefficient that they found for their calorimeter (± 0.0020225) represents the influence of the external temperature through the envelopes and packings for one minute and one degree.

Instead of a coefficient of correction thus determined, use preferably a system of correction devised by Regnault and Pfaundler. This system is superior to the preceding, as it allows consideration of all external conditions at the time of the experiment. It is evident, for example, that the evaporation of a liquid may vary in such proportions that a fixed coefficient will not always represent it.

The system of Regnault and Pfaundler does not need previous experiments nor a determined coefficient. It rests on observation of the thermometer immersed in the bath a few minutes before and after the experiment, or at the times when external influence is at its minimum or maximum. Knowing the value of these two kinds of influence, it is easy to calculate it for the whole duration of the test.

It is well to continue the observations before combustion for some five minutes. These five minutes should be preceded by at least ten minutes' immersion of the combustion chamber with agitator, so as to establish equilibrium of temperature between the cylinder and the water.

Suppose the initial correction corresponding to the first period to be zero—which is rare, it is true, but simplifies the demonstration—and that the observations have given the following data:

Initial temperature of bath............	18.460°
After 1 minute.......................	19.700
" 2 "	20.540
" 3 "	20.670
" 4 "	20.680
" 5 "	20.676
" 6 "	20.665
" 7 "	20.655
" 8 "	20.640
" 9 "	20.630
" 10 "	20.620

The combustion once commenced is continued till after the fourth minute and ends between the fourth and fifth minutes, but the equilibrium of temperature between the bath and the combustion-chamber is not established until the eighth minute, the time when the variation due to difference between them has become regular (0.010° per minute).

A table of corrections is formed as follows:

		18.460°			
1st minute....	19.700		Mean 19.080°	Difference 0.620°
2d "	.... 20.540		20.120		1.660
3d "	.... 20.670		20.605		2.145
4th "	.... 20.680		20.675		2.215
5th "	.... 20.676		20.678		2.218
6th "	.... 20.665				
7th "	.... 20.655				
8th "	.... 20.640				
9th "	.... 20.630				
10th "	.... 20.620				

The total elevation of temperature is

$$20.676 - 18.460 = 2.216°,$$

and the correction is

$$20.676 - 20.620 = 0.056° \text{ for five minutes,}$$

$$\text{or } 0.011° \text{ for one minute.}$$

Then

$$2.216 : 0.011 = 0.620 : 0.0031$$
$$2.216 : 0.011 = 1.660 : 0.0083$$
$$2.216 : 0.011 = 2.145 : 0.0107$$
$$2.216 : 0.011 = 2.215 : 0.0110$$
$$2.216 : 0.011 = 2.218 : 0.0110$$

$$\text{Total} \ldots\ldots\ldots\ldots 0.0441$$

There is then $0.0441°$ to be added to the difference, $2.216°$, increasing it to $2.260°$, which is the corrected difference of the bath temperature, from which the heat of combustion of the substance burnt in the calorimeter is calculated.

Regnault and Pfaundler's formula is

$$\Delta tn = \Delta to + K(tn - to);$$

in which

$\Delta tn =$ ascertained variation of temperature from the heating and cooling of the calorimeter for one minute;

$\Delta to =$ variation at the beginning;

$tn - to =$ loss or gain during the total time of the test;

$n =$ number of minutes of test.

Using the above numbers,

$$K = \frac{0.011}{2.216} = 0.00496.$$

It will suffice, then, to find the total loss or gain to take the sum of all the gains or losses calculated by means of the coefficient K during the whole time of the experiment.

Thus,
$$0.620 \times 0.00496 = 0.0031°,$$
$$1.660 \times 0.00496 = 0.0083°,$$
and so on.

CHAPTER IV.

CALORIMETERS WITH CONSTANT PRESSURE.

THE first calorimeters were of constant pressure; that is, the combustion was carried on at the atmospheric pressure or very near it, and did not vary from the beginning to the end of the experiment. Hence the modifications in the volume of the gases before and after combustion exercised no influence on the observed results.

Rumford, in 1814, was the first who tried to correct external influences. He employed a practical method which has often been used since, and consists in giving the calorimeter bath a temperature in the beginning of the test less than that of the room, and allowing it at the close to attain a temperature in the same proportion above that of the room. His calorimetric apparatus was composed of a copper boiler of several litres capacity, heated by an interior tube through which passed the gaseous products of the combustion. The combustible was burnt in a little burner placed under the boiler, and the air used circulated around the heater before passing to the burner, thus preventing any loss of caloric by radiation.

Dulong in 1838 used oxygen, and obtained much superior results. His calorimeter consisted of a rectangular copper box, 25 centimetres (about 10 inches) deep, 7.5 centimetres (2.9 inches) wide, and 10 centimetres (3.9 inches) long. It was closed at the upper part by a cover with a mercury seal.

The oxygen passed into the calorimeter by a copper tube opening at one of the sides of the box near the bottom. The gases of combustion were drawn into a gas-holder. The apparatus was enclosed in another likewise rectangular, in which was put 11 litres (9⅜ quarts) of water. This was the calorimetric cylinder. The water was kept in motion by an agitator.

The unit chosen by Dulong was one gram of water whose temperature was raised one degree. He corrected the temperature observed, same as Rumford, but he also noticed that this correction was correct only when the first period was equal to the second. The results obtained by Dulong in 1838 were not published till after his death, in 1843. For hydrogen and carbonic oxide they are but slightly different from the most modern determinations.

CALORIMETER OF FAVRE AND SILBERMANN.

In 1852 Favre and Silbermann published their first researches on the quantities of heat generated by chemical action and described their calorimeter.

All rapid-combustion calorimeters and all with constant pressure intended for solid bodies are copied more or less after that of Favre and Silbermann. The principle and mode of execution in their general lines are the same; the form in some details or the material employed for the combustion-chamber has been modified more or less; but the general apparatus and accessories, as well as the method, have remained as F. & S. left them. We will describe, then, this calorimeter in its details, and outline the modifications made by other experimenters.

The calorimeter called Favre and Silbermann's is composed of three concentric copper cylinders (Fig. 2, B, C, D). Cylinder B is the calorimeter cylinder; it is silver-plated and polished on the inner surface so as to lessen its emitting power; its capacity is a little over 2 litres (3½ pints), being 20 centimetres (about 8 inches) high and 12 centimetres (4¾ inches) in diameter. In the middle is placed the combustion-chamber A (Figs. 2 and 3).

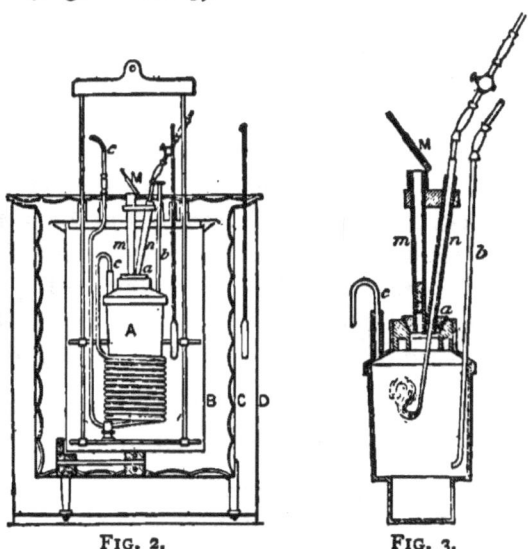

FIG. 2.   FIG. 3.

FAVRE AND SILBERMANN CALORIMETER.

The combustion-chamber is of burnished gilt copper, and is shown in Fig. 3. It is a slightly conical vessel, the large opening in which receives a stopper from which is suspended the burner made of a material suitable to that of the substance operated on. The stopper itself carries two tubes, m and n, the first being an observation tube for the combustion, and is surmounted by a mirror. M, which allows examination during the burning. The mirror receives light by the tube m, which is closed by an athermanous system of quartz, alum, and glass. The other tube, n, carries the jet for the oxygen. Tube b is closed, or removed during the test with coal, as it is of no use then. Tube c serves as the exit for the waste gases of the combustion, which pass through the coil cc (Fig. 2) before reaching the analytical apparatus. This coil is sufficient to cool the gas to the temperature of the bath. Experimenters should solder the oxygen-jet to the stopper so as to diminish the number of openings. It is also advantageous to solder the coil to the cover.

Certain fuels with very smoky flames require the addition of oxygen very near their surfaces. Scheurer-Kestner and Meunier-Dollfus employed the following arrangement (Fig. 4), a being the platinum capsule; cc', the platinum tube, which at the part c fits tight in the mouth of the oxygen-jet; b, b, b, platinum suspension-rods; d, fuel.

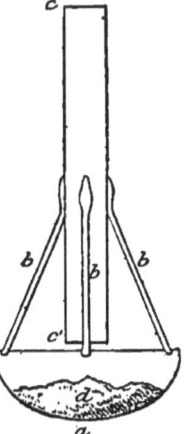

Fig. 4.

It is impossible to prevent the generation of more or less hydrocarbons and carbonic oxide. The weight of the hydrogen and carbon is determined by causing the gaseous products of combustion to pass through an organic analysis tube, after removing the water and carbonic acid. For this purpose the exit-tube c (Fig. 3) is connected by a caoutchouc tube with a Liebig apparatus, followed by a U-tube of soda-lime.

The gas-current being rather rapid, an absorption apparatus must be used, large and powerful enough to completely free the gas from the carbonic acid and water before it reaches the red-hot copper oxide. This is done by passing the gases through another U-tube smaller than the preceding, and whose weight should vary only a few milligrams. The gases thus freed pass to the tube of hot copper oxide, where the combustible gases are burnt to water and carbonic acid, which are collected and weighed as usual.

Scheurer-Kestner and Meunier-Dollfus employed a platinum combustion-tube, and prefer soda-lime as absorbent for the water after the conclusive experiments by Mulder.*

---

* Zeitschrift für analytische Chemie, I. 4.

The coal for the experiment must be in pieces; if in powder, the combustion is more difficult, unburnt gases escaping in considerable quantities, so that it is rare to obtain a complete combustion, and the cinders almost invariably contain small quantities of coke. To determine these, the capsule and tube are withdrawn from the combustion-chamber, dried, and weighed. The coke and the little soot on the sides of the capsule are burnt off by calcination in the air and a new weighing made, giving the weight of the carbon and cinder—elements which must be considered in the corrections. From half a gram to a gram of coal may be used.

When the combustion-chamber containing the weighed substance is put into the calorimeter all the parts of the apparatus are connected by caoutchouc joints and tested. A slow current of oxygen* from a gas-holder is passed through the apparatus. The combustible is ignited by a few milligrams of burning charcoal, the joint in the tube being broken for the moment, and immediately reconnected without stopping the flow of oxygen. The little glass M allows inspection of the combustion, the intensity of which can be regulated by the flow of oxygen from the gas-holder. The temperature shown by the thermometer is recorded each minute to obtain the data necessary for the correction spoken of above (pages 16 *et seq.*).

To calculate the heat-units developed by the combustion the following elements are needed:

1. Weight of the combustible used;
2. Weight of the carbon remaining in the cinders unburnt or as black;
3. Weight of the cinders;
4. Weight of hydrogen escaped unburnt;

---

* To prepare the oxygen a copper flask of one litre capacity is used, in which is placed some chlorate of potash, which is then heated by a gas flame. The gaseous current is very regular, except towards the end, when it may become tumultuous. The addition of a small percentage of black oxide of manganese promotes the regularity of the gas generation.

5. Weight of carbon escaped unburnt in the gaseous products;

6. Elevation of temperature of calorimeter bath;

7. Correction for heating and cooling caused by external influences on the calorimeter cylinder.

The combustion of the coal by this means is rarely complete; there remain variable quantities of coke mixed with the cinders formed. An uncertainty attends the calorimetric value according as the combustion was slow or rapid, since this small quantity of coke contains more or less hydrocarbons. These differences, however, apply within very close limits, so that no fear need be entertained of large errors therefrom. When a coal, in pieces, has been burnt, there remains in the capsule only a few milligrams of coke or unburnt carbon. From this we calculate the calorimetric value, using 8080 as coefficient (heat of combustion of charcoal according to Favre and Silbermann); and in using that coefficient the hydrogen which may exist in the coke is naturally neglected, but this cannot be prevented. The carbon and hydrogen of the combustible gases which escaped combustion are transformed into water and carbonic acid, and weighed as such. The hydrogen is calculated as in the free state (coefficient 34500) and the carbon as carbonic oxide (coefficient 2435).

It is evident that these are only approximations, since the hydrogen is not disengaged in a free state, but as a hydrocarbon; and its coefficient (34500) should be diminished by the heat of formation of this compound, or, in other words, by the heat of combustion of hydrogen and carbon. This correction, however, is not possible; for neither the composition nor state of molecular condensation of such hydrocarbon is known. Similarly for the carbon, and its heat of combination in the carbon compound. There are, then, some uncertainties, but not of much importance, in the determination of the heat of combustion of fuels—uncertainties which the use of the calorimetric bomb has entirely avoided.

A complete test will now be described, giving all the corrections.

Suppose one gram of dried coal in fragments is used. After combustion in the calorimeter, weigh the capsule containing the cinders.

Cinders after combustion................ 0.110 gram.
"   "   calcination in the air....... 0.100 "
Unburnt carbon remaining in cinders.... 0.010 "

Then

Coal used, dried at 100° C ,............. 1.000 gram.
Cinders................................ 0.100 "

Pure coal (cinders out)................. 0.900 "
Carbon not burnt during the experiment.. 0.010 "

There was collected from the combustion of the hydrocarbons and the carbonic oxide 0.10 gram of carbonic acid, corresponding to 0.006 of carbonic oxide (molecular ratio $11:7$); also 0.010 gram of water, corresponding to 0.0011 gram hydrogen (molecular ratio $9:1$).

Increase of temperature of the bath............................ 3.702°
Correction... 0.020
Total increase.. 3.722°

Calorimeter equiv. in water 2.114 kilos * and 3.722×2.114 = 7.8683
Unburnt carbon............................. 0.010×8.080 cal. = 0.0808
Carbonic oxide .... 0.006×2.403 " = 0.0144
Hydrogen................................... 0.0011×34.500 " = 0.0383

Total calories from 0.900 gram coal completely burnt = 8.0018

1 gram pure coal   =   8.891 calories,
1 kilogram pure coal = 8891 calories, or
1 pound   "   " = 16003.8 B. T. U.

---

* 2000 grams of water + 114 grams for value in water of calorimeter and accessories.

In this example the corrections are not very important, since they do not exceed one-half per cent. These are the ordinary conditions when the coal used is in pieces. With pulverized coal, on the contrary, the quantity of unburnt carbon and of combustible gases increases considerably and renders results less certain. The opportunity we have to weigh the cinders of each test obviates pulverization of the coal to obtain an average sample of the cinders.

Favre and Silbermann's calorimeter has been modified by Berthelot in several particulars.* He has happily modified the agitator and given it a coiled form, as shown in Fig. 5, a detailed description of which is given in his *Essai de Mécanique Chimique*, p. 145.

This agitator has the advantage over the old one of more completely mixing the water, with less force, and without accelerating evaporation. Fig. 5 shows it placed in the middle of the calorimeter. He has also replaced the gold-plated copper combustion-chamber by the glass apparatus which Alexejew used for combustibles.

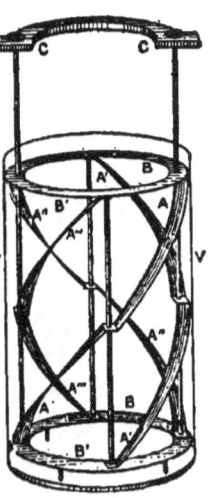

FIG. 5.

---

* The F. & S. calorimeter with all accessories and an agitator (not mechanical) costs about 500 francs ($100.00); with mechanical agitator arranged for a laboratory turbine or dynamo the cost is about 600 francs ($120.00). Berthelot's calorimetric bomb of platinum, enamelled inside and not double, costs no more, and is much preferable. A single operator can handle it, while the F. & S. apparatus requires two.

Nevertheless, the manner of working the F. & S. calorimeter is described in detail, because its use is surrounded by conditions easily realized in all countries. The calorimetric bomb requires oxygen compressed to 25 atmospheres, which cannot be obtained everywhere.

ALEXEJEW'S CALORIMETER.

The apparatus used by Alexejew was composed of a glass combustion-chamber A (Fig. 6), in which he burnt the coal previously reduced to fragments. These fragments were placed on a platinum grating in the centre of the chamber. The fuel was kindled by means of a platinum sponge placed over it, on which impinged a jet of hydrogen from the gas-holder M, opening at c, correction for which is of course made in the calculation. The grating containing the fuel was suspended from the glass rod a. As soon as the combustion was started the current of hydrogen was cut off by the cock l, and the oxygen allowed to flow in through b, the waste gases passing out through the coil. If the combustion was interrupted, it was rekindled by the hydrogen and platinum sponge. The hydrogen used was calculated in grams and multiplied by 34500. The number of calories thus obtained was deducted from that calculated from the rise in temperature of the bath. According to Alexejew, the importance of this correction never exceeded one-half per cent, and he never had to rekindle the fuel.

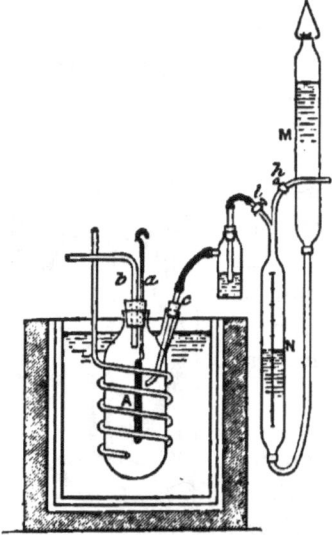

FIG. 6.—ALEXEJEW CALORIMETER.

Alexejew did not determine the unburnt gases, as experience showed they never exceeded 0.35 per cent. It is impossible, however, to determine the hydrogen of the hydrocarbons if desired, as these would be mixed with the hydrogen used for kindling, part of which may escape combustion. The kindling with hydrogen might, however, be replaced by that with carbon, as in the F. & S. apparatus.

Burning the fuel on a grating renders it impossible to weigh the cinders, and this inconvenience is of more importance as the coal is used in pieces. The use of pastilles is not possible, as they splinter in burning.

The calorimeter contained 2500 grams (5.511 lbs.) of water, a quantity somewhat larger than that usually employed, and which is based on the sensibility of the thermometer. To attain the same degree of precision it was necessary to use larger samples of fuel or else have more delicate thermometers. The water was kept in motion by the coil-agitator.

FISCHER'S CALORIMETER.

Fischer made a combustion-chamber of silver 0.940 fine, so that it would be less easily attacked by sulphur, from which the gaseous products of coal are rarely free. He drew off the waste gases at the bottom of the apparatus (Fig. 7), thus avoiding the inconvenience of exit-tubes in the cover of the combustion-chamber. The cooling coil was replaced by a flattened pipe of a certain size. A represents the combustion-chamber. The oxygen, purified by passing over potash and then dried, arrived by the tube a fastened in the tube of the cover by a caoutchouc joint, and passed by means of the platinum tube r into a crucible z of the same metal, containing one gram of the fuel. The crucible was covered by a grating, which became red-hot towards the end of the operation. This was intended to burn the waste gases, and the black deposited at the beginning. The gases flowed out at i, and after having encircled the outside

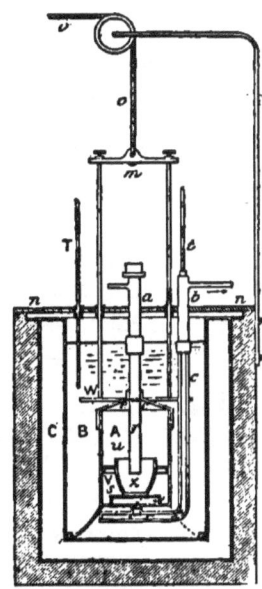

FIG. 7.—FISCHER'S CALORIMETER.

of the crucible escaped at *b*. The thermometer *t* showed whether the temperature of the gases was the same as that of the bath.

The calorimetric bath contained 1500 grams (3.3 lbs.) of water, and was protected against external influences by a wood casing, while the space *C* was filled with glass wool; but this is not necessary. *n* is a brass cover which may be dispensed with. The thermometer *T* is the calorimetric thermometer; *m* is an agitator moved by the string *o*. The value in water of the one used by Fischer was 113.5 calories. The coal was dried in nitrogen. The carbonic acid and the unburnt carbon were determined.

THOMSEN'S CALORIMETER.

This calorimeter was designed especially for tests of gases and vapors. It is not adapted to tests of solid fuels. It consisted (Fig. 8) of a calorimetric bath of thin brass, with a capacity of some 3 litres (195 cubic inches), protected from radiation by a cylindrical ebonite envelope; and a platinum balloon of half a litre (32.5 cubic inches) capacity, in which the gases were burnt, being delivered through the opening at the bottom.

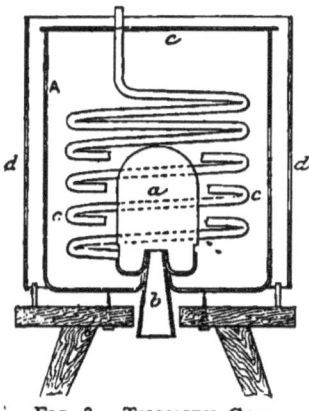

FIG. 8.—THOMSEN CALORIMETER.

The waste gases passed off through a coil, and a mechanical agitator kept the water in circulation.

The dried gas was delivered with perfect regularity from a mercury gas-holder, sufficient air or oxygen being added to render it free-burning, and enough oxygen was supplied to insure perfect combustion. This he attained by always having 40 to 50 per cent in the waste gases. The gases passed off through a carbonic acid absorbing apparatus.

To reduce to the minimum, or entirely suppress, the correction for temperature he regulated his gas-flow so that the temperature was as much higher than the air at the close of the experiment as it was lower at the beginning. This he easily did by means of his hydrogen supply. If a liquid was tested, it was vaporized and burnt in a specially devised burner which allowed complete combustion of almost all compounds not having too high a boiling-point. If too high for heat vaporization, they were carried along by a current of air, oxygen, or hydrogen, as seemed best adapted.

The water of the calorimeter being weighed, the lower portion was closed with a rubber stopper and by means of an aspirator a pressure of 8 to 12 inches of water was put on the apparatus to test the joints. When ready, the temperature of the bath and the air was noted for some minutes, the gas-holder reading taken, the burner placed in position, and the test commenced. The depression produced by the aspirator was about 0.4 inch during the whole test. The regularity of the working was shown by a gauge registering the pressure. When the temperature had reached the desired point the gas and electric current were shut off, the burner removed, and the opening closed again. The aspirator was used to draw dry air, freed from CO_2, through the apparatus to insure removal of all waste gases. The apparatus was then allowed to rest, taking the temperature at short intervals for fifteen minutes. He then had all the data required.

CARPENTER'S CALORIMETER.

Prof. R. C. Carpenter devised a calorimeter especially for coal determinations, which is a modification or extension of Thomsen's. He has used it considerably in connection with work he has been engaged on, and the results credited to him in the tables at the end of the book were obtained with it.

32 CALORIFIC POWER OF FUELS.

Fig. 9 is a sectional view of his apparatus. It consists of a combustion-cylinder, 15, with a removable bottom, 17,

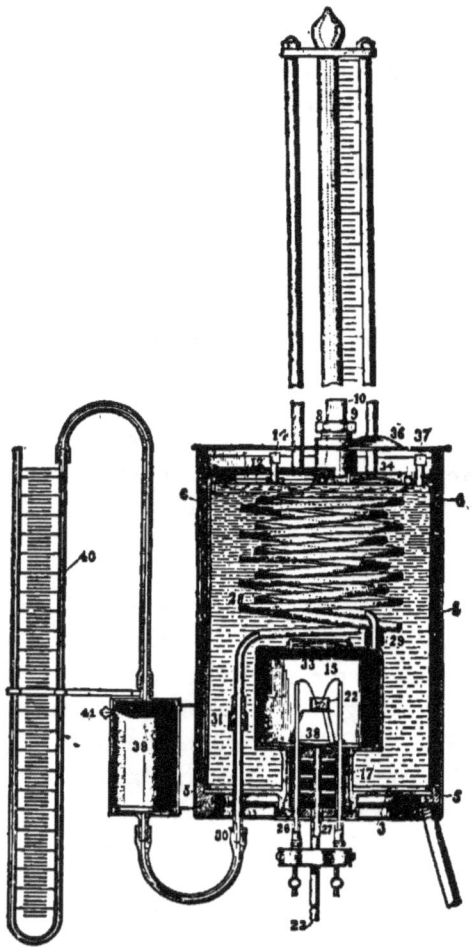

FIG. 9.—CARPENTER CALORIMETER.

through which passes the tube, 23, to supply oxygen, and also the wires, 26 and 27, to furnish electricity for the igniter. It also supports the asbestos combustion-dishes, 22, used for holding the fuel. At its top is a silver mirror, 38, to deflect the heat. The plug is made of alternate layers of asbestos and vulcanite. The products of combustion pass off through the spiral tube, 28, 29, 30, 31, which is connected with the small chamber, 39, attached to the outer case of the instrument. This chamber has a pressure-gauge, 40, and a small pinhole outlet, 41. Outside the chamber is the calorimetric bath, 1, which is connected with an open glass gauge, 9, 10. Above the water is a diaphragm, 12, used to adjust the level.

The calorimeter has an outer nickel-plated case, polished on the inside. The bath holds about 5 pounds of water, and uses about 2 grams of coal at a time. It is thus considerably larger than the bomb, and the charge being larger the time consumed by the test is longer, being some ten minutes for each gram burnt. The entire outside dimensions of the case are $9\frac{1}{2}$ inches high and 6 inches diameter.

In using the apparatus the coal is ground to a powder in a mill or mortar. The asbestos cup is heated to burn off all organic matter and weighed. The sample is then placed in it, and the whole weighed again. This gives the weight of the coal used. Place it in the combustion-chamber, raise the platinum igniting wire above the coal, make the connections with the battery, and as soon as the heat generated causes the water to rise in the glass tube turn on the oxygen, and by pulling down the wires kindle the coal. At this instant the reading on the glass scale must be taken.

By means of the glasses 33, 34, and 36 watch the progress of the combustion, and as soon as finished take the scale-reading and the time. The difference between this scale-reading and the one previously made is the "actual" scale-reading.

To correct for radiation, allow the apparatus to stand with the oxygen shut off for a length of time equal to that of the combustion, and take the scale-reading and the time. The difference between this and the "actual" reading is to be added to the "actual" for the "corrected" reading.

Now, by inspection of the calibration-curve previously prepared, at the point corresponding to the corrected scale-reading will be found the B. T. U.'s for the quantity burnt. The ash is determined by weighing the asbestos cup after the combustion.

The following shows all the calculation needed:

Weight of crucible (asbestos cup).... 1.269 grams.
" " " and coal......... 3.017 "
" " " " ash.......... 1.567 "
" " combustibles............. 1.450 "
" " ash................. .... 0.297 "
" " coal.................. .... 1.747 "

1.747 grams × 0.002205 = 0.003852 pounds.

First scale-reading....... 3.90 inches; time 2 hrs. 55 m.
Second " " .... 14.70 " " 3 " 20 "
Third " " .... 14.30 " " 3 " 45 "
"Actual" scale-reading. 14.70 − 3.90 = 10.80 inches.
Radiation correction..... 14.70 − 14.30 = .40 "

Corrected reading.................... 11.20 "

On the calibration-sheet 11.2 corresponds to 46.25 B. T. U.'s, and 46.25 B. T. U. ÷ 0.003852 = 12000 B. T. U. per pound.

All air must be removed from the water in the bath, the apparatus must work at a constant pressure, and the pressure for which it is calibrated. A pressure of 10 inches of water has been found satisfactory. Complete combustion is always attained in the asbestos cups.

It will be seen that the use of thermometers is obviated, and also all corrections but one. The apparatus is intended for ordinary every-day work, and will give good comparative results when used according to directions, which must be implicitly followed. The amount of calculation is reduced to a minimum, and there are no delicate parts requiring extra care and adjustment. For the purpose intended, it seems an advance over the others previously used, which could never give more faint approximations to correct results.

SCHWACKHÖFER'S CALORIMETER.

In 1884 Schwackhöfer published calorimetric researches on different kinds of coal, using a calorimeter in which he made

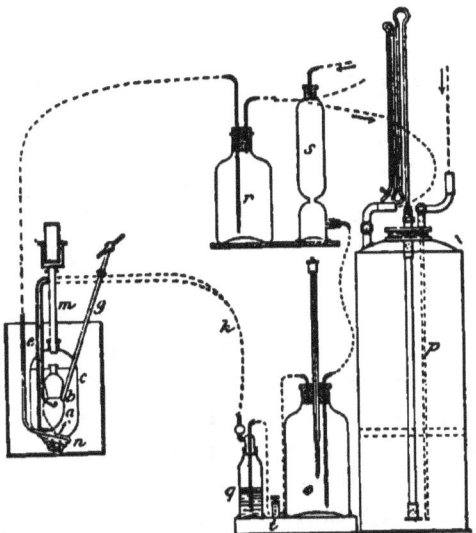

FIG. 10.—SCHWACKHÖFER CALORIMETER.

several modifications intended to render it specially applicable to such fuel.

He considered it advisable to use as much as five or six grams of coal, which is six times that generally used. He burnt at the same time and under definite conditions, shown in the sketch (Fig. 10), a certain quantity of sugar-charcoal, the combustion of which was intended to accelerate and complete that of the coal tested.

In the figure (Fig. 10) *ab* represents the combustion-chamber, *c* the calorimetric bath. Minor details of accessories, envelopes, regulators, etc., are omitted. The burner proper is of platinum and of two pieces, *a* and *b*, superimposed, the coal being placed in the lower portion, the sugar-charcoal in the upper one. All pieces of the burner may be removed for the introduction of the coal and for cleaning. The two combustibles rest on perforated plates of platinum, in which the perforations, made by a special machine, are so small that light can hardly pass through, and from which the cinders can be completely removed; the holes in the upper one are slightly larger than those of the lower. The oxygen enters through three tubes, *e*, *f*, *g*. Tubes *g* and *m* pass outside the bath, and carry mirrors to allow inspection during the burning. The waste gases pass off at the bottom through a coil *n*, and are collected in *H*. This vessel is simply to detect smoking, he having found that it happened only when the pressure was diminished at the burner, and that it could be stopped by a reinstatement of the normal pressure. *p* represents an aspirator, in which are collected the waste gases. Another one, not shown in the sketch, serves to contain the gas analyzed. Both are filled with water covered with a film of oil. The oxygen passes through a jar *s* filled with soda-lime, a bottle *o* furnished with a thermometer, a cock *t* as regulator of the flow, and one or more wash-bottles *q* containing sulphuric acid.

The calorimeter-chamber *c* contains 5200 cc. (4.6 qts.) of water. 5 or 6 grams (77 to 92.5 grains) of coal were used, with 2 to 4 grams (31 to 62 grains) of sugar-carbon of a known calorific value. The temperature of the bath rose about 10° C., and the experiment generally lasted an hour.

The sugar-carbon was first kindled in the upper part of the burner, the under portion burning first. From this sparks were thrown to the coal, and it soon kindled. The oxygen flowed in by g and e. When combustion was well under way and had reached the lower portions of the coal, g was shut off and f opened.

Schwackhöfer obtained complete combustion of the sugar-carbon and coal, with no formation of black, and no residue of coke.

The gaseous product of the combustion was generally of the following composition:

Carbonic acid............	50 to 60	per cent;
Carbonic oxide	1.2 to 0.3	" "
Oxygen...................	10 to 15	" "
Nitrogen.................	30 to 40	" "

arising principally from the fact that to keep up the normal pressure the combustion-chamber was in communication with the open air. The cinders were weighed after each test.

This apparatus should give exact results, but its use is complicated. The long duration of the test requires important corrections for influence of external heat, and it needs several thermometers.

W. THOMPSON'S CALORIMETER.

W. Thompson devised a calorimeter in which the combustion is started by a jet of oxygen, but the waste gases instead of passing through a coil bubble up through the water of the calorimetric bath. In this apparatus the uncombined gases are naturally neglected. (See Fig. 11.) It is an apparatus, as the inventor says, not intended for scientific researches, but for handy use of mechanics or " for popular use."

a is a galvanized-iron gas-holder containing oxygen; b, a stop-cock regulating the flow of water to this holder; d, stop-cock for gas; e, rubber tube; f, level-gauge; g, pressure-gauge; h, bell-glass covering the platinum crucible k, in which the coal is burnt; l is a support of earthenware suspended from the bell-glass by metal springs, and intended to insulate the crucible and prevent too quick cooling; m is a glass jar containing 2000 grams (4.4 lbs.) of water, forming the calorimetric bath. Water cannot enter the bell h while the cock j

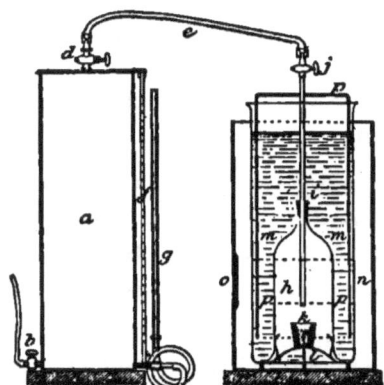

FIG. 11.—W. THOMPSON CALORIMETER.

is closed, and it is opened only when the pressure in the gas-holder is sufficient; n is a glass jar filled with water and surrounding the calorimetric jar, and p is the agitator.

One gram of fuel is put into the crucible, and on this is placed a small cotton wick impregnated with bichromate of potash. This is lighted at the instant of putting into the jar, and its combustion aided by the oxygen kindles the fuel.

This is an imperfect apparatus, and will give in most cases only unsatisfactory results. Still it is in rather common use in the shops of England, where it serves principally as a comparative measure, the errors being considered constant.

BARRUS'S CALORIMETER.

The Barrus calorimeter is a modification of the one just mentioned. While it requires considerable care in using to get correct results, yet it is one of the simplest and most inexpensive.

As described by Mr. Barrus, "it consists of a glass beaker (Fig. 12) 5 inches in diameter and 11 inches high, which can be obtained of most dealers in chemical apparatus. The combustion-chamber is of special form, and consists of a glass bell having a notched rib around the lower edge and a head just above the top, with a tube projecting a considerable distance above the upper end. The bell is 2½ inches inside diameter, 5½ inches high, and the tube above is $\frac{3}{8}$ inch inside diameter and extends beyond the bell a distance of 9 inches. The base consists of a circular plate of brass 4 inches in diameter, with three clips fastened on the upper side for holding down the combustion-chamber. The base is perforated, and the under side has three pieces of cork attached, which serve as feet. To the centre of the upper side of the plate is attached a cup for holding the platinum crucible in which the coal is burned. To the upper end of the bell, beneath the head, a hood is attached made of wire gauze, which serves to intercept the rising bubbles of gas and retard their escape from the water. The top of the tube is fitted with a cork, and through this is inserted a small glass tube which carries the oxygen to the lower part of the combustion-chamber. This tube is movable up and down, and to some extent sideways, so as to direct the current of oxygen to any part of the crucible and to adjust it to a proper distance from the burning coal."

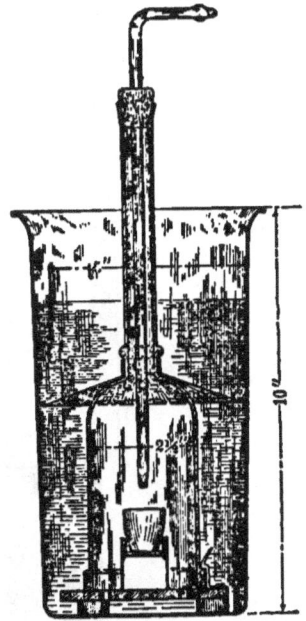

FIG. 12.—BARRUS CALORIMETER.

The method of working it can be easily seen from the description and cut. In burning very smoky coals he mixes them with a proportion of non-smoking coal of known calorific value, and when anthracite or coke is burnt he mixes it with a small portion of bituminous coal. In Mr. Barrus's hands very satisfactory results have been obtained.

HARTLEY AND JUNKER'S CALORIMETER.

Hartley's calorimeter is an apparatus of constant pressure and continued combustion. The gas measured by a meter is burnt in a Bunsen burner surrounded by a cylindrical copper

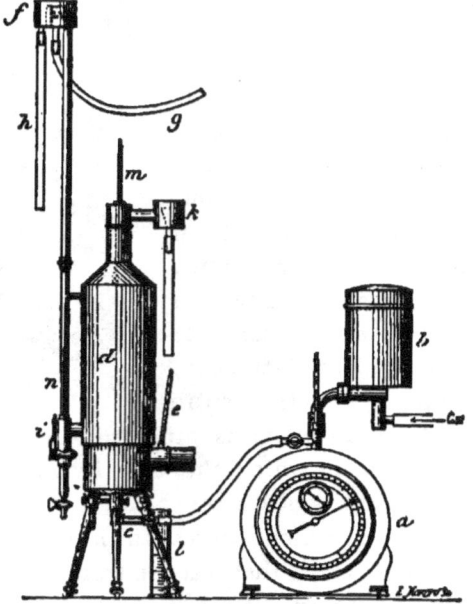

FIG. 13.—JUNKER CALORIMETER.

vessel filled with water, which is constantly renewed. The flow of liquid is such as to avoid much heating and time sufficient is used to increase the temperature so as to have a good thermometric observation. The volume or weight of the water is determined at such intervals and the thermometric readings taken often enough to obtain an average.

Hugo Junker's modification of the apparatus rendered it more exact. It has been used for some time in Germany and in the United States. It is composed (Fig. 13) of a gas-meter a, preceded by a very sensitive regulator b. On leaving the meter the gas passes to a Bunsen burner c. The products of combustion give up their heat to a calorimetric tube d, through which regularly flows a stream of water. The temperature of the gases is regulated by means of a thermometer e. In order to keep the flow of water as regular as possible, it flows from the supply-tube g into a small reservoir kept at a constant level governed by the tube h. The water passes through i to the calorimeter and escapes at k, running into the glass in which it is measured or weighed. The graduated tube l is to catch the condensed water from the interior of the calorimeter. The thermometer m shows the heat of the escaping water, and n that of the water entering the calorimeter.

To calculate the calories generated during the combustion proceed as follows:

Measure the quantity of water which runs through it in one minute, take the temperature of the two thermometers, and note the flow of gas. The heat of combustion per cubic metre of burnt gas is obtained by multiplying the volume of water flowing per minute by the difference of the two temperatures and dividing the product by the gas volume burnt per minute.

Thus:

Volume of water flowing per minute....	902.3 cc.
" " gas burnt per minute.........	2500.0 cc.
Temperature at inlet...................	13.1° C.
" " outlet...................	27.5° C.

$$Q = \frac{902.3 \times (27.5 - 13.1)}{2.5} = 5196 \text{ calories.}$$

The gas tested has a value of 5196 calories per cubic metre.

Since the calorie is 3.968 times the B. T. U., and the cubic metre is 35.316 times the cubic foot, multiplying the calories per cubic metre by $\dfrac{3.968}{35.316} = 0.11235$ will give B. T. U.'s per cubic foot.

Multiplying, then,

$$5196 \times 0.11235 = 583.8 \text{ B. T. U.'s per cubic foot.}$$

The above example considered the volume of the water. It is sometimes advisable to consider the weight instead. The following example illustrates this:

Weight of water used during the test.... 2000 grams.
Volume of gas burnt.................... 7.23 litres.
Temperature at inlet................... 14.4° C.
"   " outlet................... 36.5° C.

Then

$$Q = \frac{2000 \times (36.5 - 14.4)}{7.23} = 6102 \text{ calories per cubic metre,}$$

and

$$6102 \times 0.11235 = 685.6 \text{ B. T. U. per cubic foot.}$$

Two causes of error may occur. It is not certain that the combustion of the gas in the burner is regular; indications by gas-meters are not always very sure, the start being capricious. But these do not have much weight in its use for industrial purposes, for which it is chiefly designed. The results are very near those obtained by other methods. Stohmann, whose competence in such matters is universally recognized, says they give good results.

Bueb-Dessau, to prove the calorimeter, burnt hydrogen prepared by electrical decomposition, and obtained after corrections for thermometer and barometer 34150 calories per kilogram—a difference of 350 calories from the usual number, 34500, or only 9 thousandths.

Prof. Jacobus has determined that there is a constant error due to neglect of latent heat of moisture in products of combustion of —2 per cent in the determinations with this apparatus; otherwise it is very satisfactory.

LEWIS THOMPSON'S CALORIMETER.

Lewis Thompson's calorimeter has been used in England for some time. It gives only approximate results, but as the errors are of the same kind in each case, the results are comparable, and it has been found serviceable in industrial works where quick and comparative observations are required.

The apparatus (Fig. 14) is composed of a glass calorimeter-bath H containing water, a copper cylinder E in which the

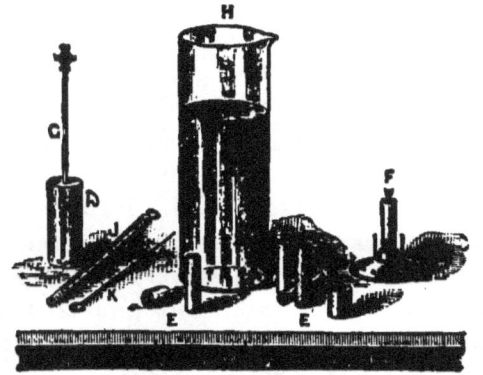

FIG. 14.—L. THOMPSON CALORIMETER.   FIG. 15.—CALORIMETER IN ACTION.

mixture of coal and potassa chlorate is placed, and surmounted by the nitrate of lead fuse F. Enclosing this cylinder is a bell D, having a tube C carrying a stop-cock. The cock is closed before putting it in position in the water. K is a cleaner for the tube C, and J is a thermometer.

The fuze is lighted, and the whole quickly put in the jar of water. The mixture of combustible and potassium chlorate soon ignites and burns, all the gases generated being forced out at the bottom of the bell through the perforations, and bubble up through the liquid. After the combustion is finished the temperature is taken and the heat-units calculated.

From 8 to 10 parts of oxidizing mixture is recommended for one of coal; but if the coal is very rich this must be increased to 11 parts, calculated on the crude coal. With pure coal, cinders out, the extreme limits are 11 and 14 parts. It would probably increase the accuracy of the method, if the same quantity of oxidizing mixture was employed, whatever the kind of coal used, and to mix with it inert substances, as silica or ground porcelain, in quantity varying with the richness of the coal.

Scheurer-Kestner tested this apparatus very carefully, using a great variety of fuels whose heats had been previously ascertained by means of Favre and Silbermann's calorimeter. He found some 15 per cent deficit in the figures, and after correcting by this amount the results varied only a few per cent from those actually obtained. In thirty different kinds of coal tested the average was 1.8 per cent too low.

The use of this calorimeter requires some skill. Its imperfect insulation requires prompt reading and rapid combustion. Care must be taken to work at temperatures very close to that of the room, as the calorimetric bath is not protected. The proportions of the mixture used vary, not only with each kind of coal, but for each sample, on account of the proportions of cinders. Fat coals require more oxidizer than lean coals, as it is evident an increase in quantity of cinders should require a decrease in oxidizer. But in changing the proportions of oxidizer a certain difference in elevation of temperature is necessarily produced by the heat of solution of the salts left after the combustion. These various causes render its working rather delicate, and always uncertain.

CHAPTER V.

CALORIMETERS WITH CONSTANT VOLUME.

THE results obtained with a calorimeter of constant volume are not exactly the same as those obtained with one of constant pressure; but for solid or liquid substances the difference is too small to consider, since the volume, as well as that of the water produced, is inconsiderable in relation to the volume of gas employed. As regards the correction for contraction and expansion of the gases, they also are inconsiderable.

In his *Traité de Mécanique* Berthelot has shown that the heat generated by a reaction between gases at constant pressure is equal to the heat of combination at constant volume at any temperature whatever, increased by the preceding product counting from absolute zero; and he gives the following formula for passing from one system to the other:

$$QT_p = QT_v + 0.5424(N - N') + 0.002(N - N')t,$$

QT_p being the heat generated by the reaction at constant pressure, and at the temperature T counting from ordinary zero; QT_v, the heat generated by the reaction at same temperature and constant volume; N, the number of units of molecular volume occupied by the components, these being taken according to usage equal to 22.32 litres under normal pressure at 0°; N', the corresponding number of units of molecular volume occupied by the product of the reaction.

As example, take the combustion of carbonic oxide at 15°. Then we have

$$CO + O = CO^2 \text{ generates at constant volume 68 calories.}*$$

---
* These numbers refer to molecular weights.

To pass from this to the heat given off under constant pressure, observe that CO occupies a unit of volume and O a half unit. Then

$$N = 1\tfrac{1}{2}.$$

CO_2 occupies a unit of volume and

$$N' = 1.$$
$$\text{Hence} \quad N - N' = \tfrac{1}{2}.$$

At 0° there would be, then, for the difference between the heat of combustion at constant pressure and that at constant volume,

$$+ 0.542 \times \tfrac{1}{2} = + 0.271 \text{ calories}.$$

At $+ 15°$ add to this $+ 0.015$, which increases the correction then to 0.286. The heat of combustion of carbonic oxide at constant pressure and 15° is then $+ 68.29$ calories.

With a solid or liquid, this volume in relation to those of the gases formed may be practically neglected, the same as with the water; all reduce then to the contraction and expansion of the gases. Thus, for naphthalin, this correction does not exceed 8.8 in 9692 calories—less than 0.1 per cent.

In case of solids or liquids with unknown molecular weight, as with fuels generally, this difference may still be approximately calculated, as it is sufficient to know the volume of oxygen used in the combustion and that of the gases produced.

The first calorimeter of constant volume in date is that of Thomas Andrews, who in 1848 published results obtained with a closed calorimeter. The calorimeter was not applicable to solids or liquids; the combustion of the gases was conducted as in a eudiometer, but he did not take all the precautions necessary to be certain of complete combustion.

Nevertheless, the results obtained for certain gases are remarkable, considering the elementary character of his apparatus and working. The combustion of solids, on the contrary, gave worthless results.

The calorimetric bomb of Berthelot and Vielle seems able to replace advantageously all the other calorimeters as much by its convenience as by its certainty of results.

Aimé Witz made certain changes in the bomb designed to facilitate its use, and devised his "calorimetric eudiometer," in which only gases can be burnt. The apparatus is more convenient than the bomb, but this convenience has been gained at a sacrifice of precision. It is more an instrument for practical use than a scientific calorimeter, but may be useful within narrow limits.

ANDREWS' CALORIMETER.

In 1848 Andrews published his labors on the heat of combustion of bodies, and notably on that disengaged by combustion of different gases. He used a calorimeter of constant volume, in which the combustion-chamber was a copper cylinder (Fig. 16) weighing 170 grams (6 ounces), of 380 cubic centimetres (about 23½ cubic inches) capacity, and capable of resisting the pressure exerted by the combustion of the same volume of olefiant gas (C_2H_4) with oxygen.

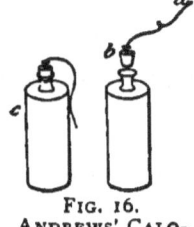

FIG. 16. ANDREWS' CALORIMETER.

At the upper part, the cylinder had a small conical tube closed by means of a perfect-fitting stopper b. A silver wire a was fixed in this stopper, and to this was soldered a very fine platinum wire for igniting the gases by a galvanic current. The mixture of gases was prepared as for eudiometric analysis.

The combustion-chamber was entirely submerged in a glass cylinder filled with water, of which the temperature is regulated so as to compensate approximately for the probable use, and thus avoid corrections for influence of external air. This cylinder was put into another, also of glass. A rotary motion imparted to the cylinder aided circulation in the liquid during combustion, which usually lasted thirty-five seconds.

Andrews also applied his calorimeter to combustion of solids, but judging from the low results he did not have perfect combustion. The results obtained with some of the gases, on the contrary, are quite reliable, notwithstanding the imperfections of the apparatus.

CALORIMETRIC BOMB OF BERTHELOT AND VIELLE.

Of all the calorimeters known to-day, the calorimetric bomb of Berthelot is that which offers the most advantages, as much from its ease of operation as from the precision of its results. Only one operator is needed; the combustion is perfect; the gaseous products need not be analyzed to determine the combustible substance; no weight save that of the substance used is needed; and it is as applicable to solids and liquids as to gases.

True, its use requires oxygen under high pressure; but this pressure (25 atmospheres) may be readily obtained with a compression-pump, which is easily procured; and at the present time oxygen may be bought sufficiently compressed for the purpose. Berthelot states that as much as 5 or even 10 per cent of nitrogen is allowable, but that the latter limit must not be exceeded.

Mahler used compressed oxygen, and obtained good results with that bought in the Paris market. This gas is furnished in steel tubes and under 120 atmospheres pressure. The cylinders contain sufficient gas to make a large number of experiments before the pressure falls too low, i.e., below 25 atmospheres.

Fig. 17 shows the bomb adjusted ready to place in the calorimeter. Full details of the construction will be found in Berthelot and Vielle's treatise, *Sur la force des métiers explosives*, vol. 1, p. 245.

Fig. 21 shows the arrangement adopted by Berthelot to burn solids. The cylinder (Fig. 18) is lined with platinum, and constructed so as to resist a pressure of 200 to 300 atmospheres. It is furnished with a tight-fitting head (Fig. 17) fastened exteriorly by a piece of steel (Fig. 19), clamped on the external face of the bomb by a screw-clamp (Fig. 20), which does not form a part of the apparatus as immersed.

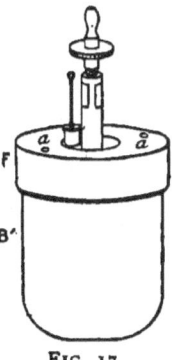

FIG. 17.

The sealing of the bomb results from the adherence of the margin of the head *BB* (Fig. 21), and the interior of the cylinder, and also between the platinum of the head and the platinum of the cylinder. Berthelot makes the joint

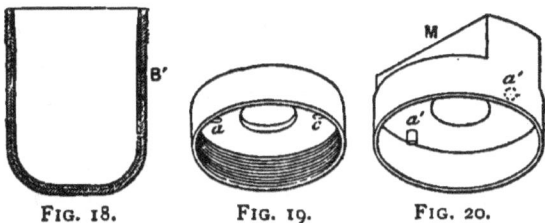

FIG. 18.   FIG. 19.   FIG. 20.

tight with a smearing of vaseline around the opening, being careful not to have a trace on the inside. If no bubbles escape on putting it into the calorimetric bath, the joints are tight.

The cover is pierced at the centre with a small hole, in which is fitted a tube formed of a hollow screw acting as a cock, and itself provided at the upper end with a circular head. The electric ignition is produced by a platinum wire fitting in an opening of the removable conical cover E. This is prepared (Fig. 21) in advance, and is covered with a layer of gum lac applied in a strong alcoholic solution. When the first coat is dry, a second one is put on and dried in a stove. Berthelot says that the combination of these two coatings, one elastic and soft, the other hard and brittle, resists very well the enormous pressure on the cone. This cone, lightly greased, is put into the conical opening in the bomb cover, and screwed up tight by means of a nut. It is well to protect the base of the cone by a film of mica.

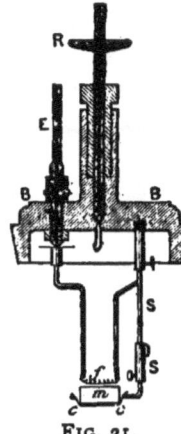

FIG. 21.

An electric current passed through E (Fig. 21) reddens the spiral of very thin iron wire f placed between the platinum wires and one of the supports SS of the capsule cc containing the substance m. This iron wire soon burns and kindles the combustible.

Fig. 22 gives a general and complete internal view.

The iron spiral is formed of an iron wire $\frac{1}{10}$ millimetre (0.004 inch) thick, rolled up on a spindle. The wire may be weighed, or by using the same length of wire always have the same weight.

The spiral is attached on one side to the cone, and on the other side by means of a platinum wire to the platinum supporting the fuel, taking care that the iron has no straight portions. The support of the capsule or platinum-foil is then fixed in the cover, by aid of the screw, arranging it so that the spiral is directly over the combustible used. The cover is put on, turning it gently to make the contact more perfect. The nut is tightened and the wire carefully screwed up, always using wooden tongs to prevent injuring the bomb.

The form of the bomb is such as permits filling the calorimeter with the smallest possible quantity of water—a necessary condition that the temperature, and consequently the precision, attain a high degree. For solids and also for coal Berthelot uses bombs containing 400 to 600 cubic centimetres (24 to 37 cubic inches), placed in a calorimeter of 2000 grams (4.4 lbs.) of water.

To determine the heat of combustion of coal, for instance,

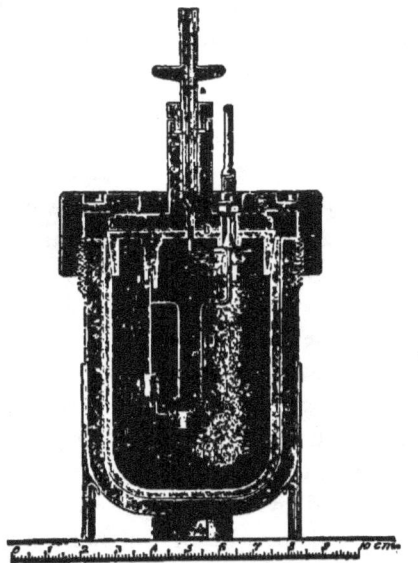

Fig. 22.—Berthelot Bomb.

it must be previously reduced to powder in order to have a sample whose cinder is known. As all kinds of coal do not burn completely in this state, they are formed into pastilles,* which are weighed and burnt. They are put on a platinum grating or foil, placed on the support SS (Fig. 21), over

---

* We obtain very resisting pastilles or briquettes from fat coals by simple compression in a pastille or suppository mould such as used by druggists. With lean coals, or anthracite, the pastilles are too friable and burn incompletely. This is easily remedied by mixing with a small quantity of silicate of soda solution. Several of them should be made at a time, the cinders of some being determined to obtain a mean and the others burnt in the bomb. They may contain about 1 gram of pure coal.

which and in contact with it is the iron spiral. At the instant of lighting a slight noise is made, and soon the thermometer begins to rise, showing that the combustion is proceeding.

Compressed oxygen may be introduced either by a pump drawing the gas from a holder or by using a compressed-gas cylinder. In both cases the gas is used without drying, if the combustible contains hydrogen in quantity enough to saturate the gases formed with water produced by its combustion. But if, on the contrary, the combustible has little or no hydrogen, like wood-charcoal for instance, it is not immaterial whether the oxygen be dry or not. In this case it is well to use the oxygen moist, or to put a little water in the bomb on the internal walls. By this means a correction for heat of vaporization of water formed by the combustion is obviated.

Oxygen compressed to 120 atmospheres is nearly dry. Berthelot observes: "The oxygen is, in short, actually or nearly dry, and if it contains aqueous vapor the tension is reduced to one fourth or one fifth on account of the change in volume of the gas during its passage through the bomb. It may be nearly nullified by the cold produced at the instant of filling the bomb. This admitted, we shall have to account in most combustions for the evaporation of the water produced in the bomb; and this is from 2 to 3.5 calories in a bomb of $\frac{1}{3}$ litre (about 0.6 pint), or 5 to 6 calories in a bomb of 600 to 700 cubic centimetres (37 to 43 cubic inches). These are rather small quantities, it is true; but while they can be neglected in industrial tests, they cannot in rigorously scientific investigations. This correction may, however, be neutralized by putting into the bomb 4 or 5 cc. of water, which should be considered in the calculations.

When oxygen not previously compressed is used and forced in by a pump, Berthelot recommends passing the gas through a large red-hot copper tube filled with oxide of the same metal, so as to burn any oil which may have been taken from the pump.

Operation.—At the laboratory of the College of France the successive operations are as follows:

1. Light the fire to heat the oxygen red-hot;
2. While the gas-holder is filling with oxygen, the fuel is dried;
3. Weigh the fuel;
4. Place the fuel in the bomb;
5. Grease the cover slightly; tighten with the screw;
6. Begin to compress the oxygen by forcing the air out with a few strokes of the piston; pump slowly to prevent heating the pump;
7. Close the stop-cock of the pump; break the connection with the bomb, extinguish the fire, and replace the bomb on its support so as to carry it to the calorimeter room;
8. Pour the water into the calorimetric bath.

The apparatus is allowed to come to equilibrium, and the readings of the thermometer taken for five minutes. The iron coil is then heated by the electric current from a small bichromate battery. It takes fire and kindles the combustible, which generally burns without smoke or producing any carbonic oxide, as Berthelot has shown.*

The water condensed from the combustion contains small quantities of nitric acid, showing imperfectly purified gas. This may be determined by titration, if accurate results are sought, and calculated 0.227 calories per gram of HNO_3. The correction will be very small. A correction for the iron used may be made at the rate of 1.65 calories per gram, this being the heat of formation of the magnetic oxide.

---

* With very fat coals it sometimes happens after a combustion that the platinum shows a black or brown mark, indicating a slight deposit of black or tar which has escaped combustion. Occasionally, also, a trace of tar is found at the bottom of the bomb. These may be prevented by using a grating or perforated plate instead of the foil. This detail must be attended to with a new coal.

With substances containing nitrogen and sulphur, such as coal, the corrections are more complicated, as a larger quantity of nitric acid is formed and the sulphur forms sulphuric acid. If exactness is sought, it will not be sufficient to make a volumetric test: the sulphuric acid must be determined separately. Generally, however, this estimation may be dispensed with, if for technical purposes only. When, on the contrary, absolutely correct figures are desired, both acids must be considered. In the calculation the nitric acid is reckoned as 0.227 calorie per gram and the sulphuric acid as 1.44 calories per gram.

But these two corrections are really unimportant even with coal, as it contains usually only about 1 per cent of nitrogen or sulphur. One per cent of nitrogen represents $4\frac{1}{2}$ per cent of HNO_3, or 10 calories; one per cent of sulphur represents 3 per cent of H_2SO_4, or 43 calories,—both quite small compared with 7000 to 8000 calories.

Below will be found the details of a complete combustion taken from Berthelot's work.

HEAT OF COMBUSTION OF CARBON.

The wood charcoal, purified by chlorine at red heat to remove all traces of hydrogen (Favre and Silbermann's method), is dried at 120° to 140° C. (248° to 284° F.), then weighed in a closed tube after cooling in a sulphuric acid desiccator.

0.437 gram carbon; cinders, 0.0028 gram (0.66 per cent); real carbon, 0.4342 gram.

Preliminary Period.

0 minute	17.360°	3d minute	17.360°
1st "	17.360	4th "	17.360
2d "	17.360		

COMBUSTION.

5th minute	18.500°	7th minute	18.820°
6th "	18.782	8th "	18.818

SUBSEQUENT PERIOD.

9th minute	18.810°	12th minute	18.785°
10th "	18.802	13th "	18.775
11th "	18.795	14th "	18.768

Initial cooling per minute,

$$\Delta t_0 = 0.00°.$$

Final cooling per minute,

$$\Delta t n = + 0.008°.$$

Correction for cooling,

$$\Delta t = + 0.056°.$$

Variation of temperature, uncorrected,

$$18.818° - 17.360° = 1.438°.$$

Value of corrected temperature,

$$1.438° + 0.056° = 1.484°.$$

Value in water of the calorimeter (including oxygen),

$$m = 2398.4.$$

Weight of acid formed;

$$HNO_3 = 5 \text{ cc. of } \tfrac{1}{20} \text{ normal KHO} = 0.0173 \text{ gram.}$$

Total heat observed, $q_t = 3.5562$ calories.
Heat of iron coil, 22.4 ⎱
" " 0.173 HNO$_3$, 3.9 ⎰ $q_1 = 0.0263$ "

Real heat due to the carbon, 3.5299 "

or for one gram, $\dfrac{3.5299}{0.4342} = 8.1296$ calories,

or per kilogram, 8129.6 calories,

or 14871.0 B. T. U. per pound.

CHAPTER VI.

THE CALORIMETRIC BOMB ADAPTED TO INDUSTRIAL USE BY MAHLER.

THE calorimetric bomb of Berthelot costs considerably more than can be paid by an industrial laboratory, owing to its large amount of platinum. Mahler replaced the interior platinum of the bomb by an enamel deposited on the steel. The description given by him in his paper before the *Société d'Encouragement de Paris*, in June, 1892, is as follows:

The apparatus is shown in Fig. 23. It consists essentially of a steel shell, B, capable of resisting 50 atmospheres

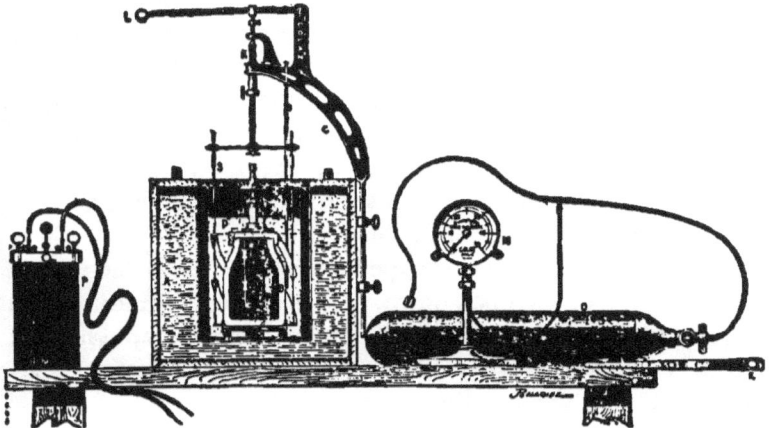

FIG. 23.—MAHLER CALORIMETER.

and 22 per cent elongation. This quality was carefully chosen, not only on account of the pressure it must stand, but also as it aids the enameling. The metal is very pure, containing but little phosphorus or sulphur. Tensile strength tests are the best criterion of quality.

It has a capacity of 654 cc. (40 cubic inches) at 15° C. It is gauged with a balance showing $\frac{1}{50000}$. The total weight is about 4 kilograms (8.8 lbs.) with the accessories.* The metal of the walls is 8 millimetres (about 0.3 inch).

The capacity is greater than Berthelot's, and has the advantage of insuring perfect combustion of carbon in all cases, due to a certain excess of oxygen, even when the purity of this gas as bought is not quite satisfactory. Besides, it is designed to study all industrial gases, even those containing a large percentage of inert gas; hence it must be able to use a sufficiently large quantity to generate the required temperature. The contraction at the top aids in enameling.

The shell is nickeled on the outside, while internally it has a coating of white enamel, resisting corrosion and oxidizing action of the combustion.† It does not, however, offer resistance to the heat, being very thin, and it weighs only about 20 grams (308 grains).

It is closed by an iron stopper made tight by a lead washer (P, Fig. 24) and clamped down. This carries a conical-seated stop-cock, R, of fine nickel—a metal almost unoxidizable. An electrode well insulated and reaching the interior by a platinum wire runs through the stopper.

Fig. 24 shows most of the details.

Another platinum wire, also fixed on the cover, supports the platinum disk or foil on which the fuel is placed.

The calorimeter, the non-conducting material, the support for the shell in the water, and the agitator differ in numerous details from those of Berthelot, and are much cheaper.

---

* Slight modifications have been made in the dimensions of the metal of the bombs made lately by Golaz.

† Prof. W. O. Atwater finds that the enamel chips off in time, and that after about 300 combustions it requires re-enameling. Hempel for coal determinations uses one without any inside enamel.

The calorimeter is of thin brass, and is quite large on account of the size of the combustion-chamber. It contains 2200 grams (4.85 lbs.) of water, thus eliminating the causes of error due to the loss of a few drops by evaporation.* The agitator of Berthelot is supplanted by a very simple and gentle cinematic combination called a drill movement, and which can be worked without fatigue. The source of electricity is a Trouvé bichromate pile (P, Fig. 23) of 10 volts and 2 amperes.

The oxygen used is that furnished by the *Compagnie Continentale d'Oxygène*. This company supplies oxygen free from CO_2, but containing from 5 to 10 per cent of nitrogen. This means of supply simplifies the manipulation; it also obviates the introduction of grease, as happens with oxygen compressed by a pump in the laboratory.†

The cylinders vary in size, and contain gas at a pressure of 120 atmospheres. The average content is about 1200 litres (about 40 cubic feet) compressed. They

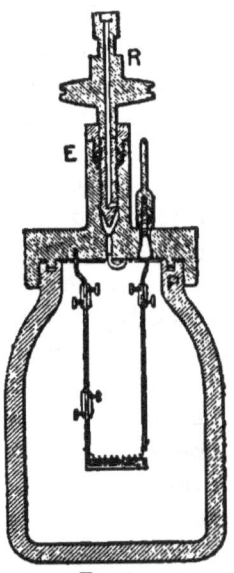

FIG. 24.

have a uniform top, and hence the copper pipe connecting the bomb with the manometer and the cylinder, once adjusted, will fit all of them.

The method of working is very simple.

Weigh 1 gram of the substance to be tested in the capsule. Fasten a small weighed iron wire (English gauge 26 or 30) to the electrode and to the support of the capsule. Put the end in the bomb and fasten in the cover, which should be held in a vise. Put the conical stop-cock in connection with the oxygen cylinder, and open it carefully so as to allow suffi-

---

* The evaporation never exceeds a gram per hour.
† This gas is also compressed by pumps at the works.

cient oxygen to pass in for the required pressure. Close the cock of the oxygen cylinder, carefully close the conical cock, and break the connection between the bomb and the oxygen cylinder. The substance, especially if coal, must not be too fine, and the oxygen must flow in very slowly to avoid blowing any of it from the capsule.

The bomb thus prepared is placed in the calorimeter, and the thermometer and agitator adjusted. Pour in the previously weighed water, agitate a few minutes to restore equilibrium of temperature, and commence the observations.

The experimenter notes the temperature minute by minute for four or five minutes, and determines the rate of the thermometer before the combustion. Then he joins the electrodes, and the combustion begins immediately, almost instantaneously; but the transmission of heat to the calorimeter takes some time.

The temperature is taken one-half minute after kindling, then at the end of the minute, then at each minute to the time when the thermometer begins to lower regularly. This is the maximum. The observations are continued for a few minutes more to ascertain the rate of fall of temperature.

We now have all the elements needed for the calculation, and particularly for the single correction necessary to make under the circumstances. This is the correction for loss of heat before reaching the maximum temperature, which is quite small considering the short time and the large mass involved.

It is not necessary to use the corrections of Regnault and Pfaundler with this apparatus. Newton's law of cooling gives sufficiently accurate results, even in rigorous investigations. Special experiments made to determine the rate of cooling of the water in the calorimeter, when the apparatus was set up as usual, showed that the correction may be regarded as following a simple law, but between comparatively large limits, even under a variation of several hundred grams in amount of water used.

The law* is

1. The decrease in temperature observed after the maximum represents the loss of heat of the calorimeter before the maximum and for a certain minute, with the condition that the mean temperature of this minute does not differ more than one degree from the maximum.

2. If the temperature considered differs more than one degree but less than two degrees from the maximum, the number representing the rate of decrease dimminished by $0.005°$ will be the correction.

The two preceding remarks suffice in all cases with Mahler's apparatus. The variation of heat in the first half-minute after kindling may also be corrected by the same law.

The agitator must be worked continually during the experiment, being careful of the thermometer.

When through, the conical valve is opened and then the bomb. Wash the inside with a little distilled water to collect the acids formed. The proportion of acids carried away by the escaping oxygen at the opening may be neglected. Determine the acids volumetrically.

When experimenting with substances low in hydrogen and incapable of furnishing sufficient water to form nitric acid, it is advisable to put a little water in the bomb, or hyponitric acid would be formed.

All the data being obtained, we proceed to the calculation of the calorific power Q.

Let Δ be the observed difference of temperature;
- a, the correction for cooling;
- P, the weight of water in the calorimeter;
- P', the equivalent in water of the bomb and accessories;

---

* It is evident that the rule must be modified for apparatus notably different from that used by Mahler.

p, the weight of the nitric acid, HNO_3;
p', the weight of the iron;
0.23 calorie, the heat of formation of 1 gram of nitric acid; and 1.6 calories, the heat of combustion of 1 gram of iron.

We then have

$$Q = (\Delta + a)(P + P') - (0.23p + 1.6p').$$

In testing coal in this manner the small amount of sulphuric acid formed will be reckoned as nitric acid without serious error, as it will be very small. The heat of the reaction is 1.44 calories per gram of H_2SO_4 formed.

The above details apply to liquids as well as solids. Heavy liquids, such as the heavy oils, tars, etc., are weighed directly into the capsule; but light, easily vaporized liquids must be placed in pointed glass bulbs. These are put into the capsule, and just before closing the bomb are broken to allow access of the oxygen to the liquid. An almost perfect combustion is obtained in operating with a great variety of materials, nothing but cinders remaining.

To determine the calorific power of gases the exact content of the bomb must be known. Fill it first with gas. Then work the air-pump to reduce the pressure to several millimetres of mercury, and then fill the bomb again with gas, under atmospheric pressure and at the laboratory temperature. The bomb may then be considered full of pure gas.

The method of working with gases is the same as with solids or liquids. The operator must not forget the need of preventing too great dilution with oxygen, as then the mixture will cease to be combustible. With illuminating gas 5 atmospheres of oxygen is sufficient, and with producer gas only one-half atmosphere, as shown by the mercury gauge, is needed.

The gases to be burnt are kept in gas-holders over water saturated with gas, or over salt water, according to circumstances, and are saturated with aqueous vapor when they enter the bomb. From the calorific capacity of the different parts we obtain that of the whole, the glass and enamel being omitted.

Soft steel.......3945 grams. $3945 \times 0.1097 = 432.76$
Brass........... 545   " $545 \times 0.093 = 50.68$
Mercury, platinum, and lead 72 ' $72 \times 0.03 = 2.16$

                              Sum............... 485.60 grams.

The coefficient 0.1097 is the one adopted by the College of France, from Berthelot and Vielle's experiments, for a steel of similar quality. We have given above (page 14) the calculations relative to the valuation in water. By direct method of mixing water of different temperatures Mahler found the equivalent to be 470 and 484, and assumed the mean 481.

By the method of burning a body of known composition and heat of combustion he obtained with naphthalin 9688 calories—within $\frac{1}{2000}$ of that given by Berthelot (9692).

The equivalent in water may also be obtained by burning 1 gram of known composition and heat of combustion—naphthalin for instance.* We may also, after Berthelot, burn a substance of fixed composition at two trials with different weights of water in the calorimeter. Two equations are thus formed, from which the heat of combustion of the body used is eliminated, and the heat sought obtained.

In using naphthalin care must be taken to weigh it only after being gently fused in the capsule. It is so light that if not agglomerated some would be blown away by the oxygen. In practice the tests are made rapidly. The water equivalent once determined may be verified by combustion of cane-

---

* This practical method has the advantage of automatically eliminating causes of error.

sugar ($C_{11}H_{11}O_{11}$), for which Berthelot and Vielle found 3961.7 calories. (Use 2 grams for a combustion.)

Examples of Calculations.

Mahler gives several types of calculations from his notes, so as to show the different circumstances which may occur.

A. **Colza Oil.**—Elementary analysis showed—

Carbon............................	77.182	per cent.
Hydrogen	11.711	" "
Oxygen and nitrogen............	11.107	" "
	100.000	" "

Weight taken, 1 gram. Calorimeter contained 2200 grams water. Equivalent in water of bomb, etc., 481 grams. Pressure of oxygen, 25 atmospheres.

The apparatus prepared as above was allowed to rest a few minutes to gain equilibrium of temperature. Then commenced noting the temperatures.

PRELIMINARY PERIOD.

0 minute............	10.23°	3 minutes............	10.24°
1 "	10.23	4 "	10.25
2 minutes............	10.24	5 "	10.25

Rate of variation,

$$a_0 = \frac{10.25 - 10.23}{5} = 0.004°.$$

The electrodes are connected and the combustion begins.

COMBUSTION PERIOD.

5½ minutes	10.80°	7 minutes..	13.79°
6 "	12.90	8 " ..	13.84 maximum.*

---

* Prof. Jacobus recommends plotting the temperatures and using, not the maximum, but the one at the instant the curve of cooling becomes a straight line. The difference is slight, but important in some cases.

Period after Maximum.

9 minutes......... 13.82°	12 minutes .... 13.79°
10 " 13.81	13 " 13.78
11 " 13.80	

Rate of variation after maximum is

$$a_t = \frac{13.84 - 13.78}{5} = 0.012°.$$

The thermometer observations now stopped.
The gross variation in temperature was

$$13.84 - 10.25 = 3.59°.$$

The corrections are as follows:

The system lost during the minutes (7, 8) and (6, 7) a quantity of heat corresponding to $2a_t$.

$$2a_t = 0.012 \times 2 = 0.024°.$$

In the half-minute (5½, 6) it lost

$$\tfrac{1}{2}(a_t - 0.005) = 0.0035°.$$

But during the half-minute (5, 5½) it gained

$$\tfrac{1}{2}a_0 = \frac{0.004}{2} = 0.002°.$$

Consequently, the loss for the minutes (5, 6) is

$$0.0035 - 0.002 = 0.0015°.$$

So that the system had lost, before reaching the maximum temperature,

$$0.024 + 0.0015 = 0.0255,$$

which must be added to the 3.59° already found, making the variation in temperature 3.615°, neglecting the 4th decimal.

The quantity of heat observed, then, is

$$Q = (2200 + 481)3.615 = 2681 \times 3.615 = 9.6918 \text{ calories.}$$

From this number must be subtracted—

1. The heat of formation of the 0.13 gram of HNO_3, ... $0.13 \times 0.23 = 0.0299$
2. The heat of combustion of 0.025 gram of iron wire............ $0.025 \times 1.6 = 0.04$

                    Total subtraction.................. 0.0699

The final result is, then,

$$9.6918 - 0.0699 = 9.6219 \text{ calories,}$$

or for 1 kilogram 9621.9 calories, equivalent to 17319.4 B.T.U.

TECHNICAL EXAMINATION OF COAL.

The coal taken was a sample of Nixon's coal from South Wales.

Preliminary Period.		Combustion.		After Combustion.	
minutes.	degrees.	minutes.	degrees.	minutes.	degrees.
0	15.20	3½	16.60	7	18.32
1	15.20	4	17.92	8	18.30
2	15.20	5	18.32	9	18.30
3	15.20	6	18.34	10	18.30
			maximum	11	18.26
$a_0 = 0$		oxygen pressure 25 atmospheres		$at = \dfrac{18.34 - 18.26}{5} = 0.016°$	

Difference of gross temperature .... 3.140°
Correction (4, 5) (5, 6) 0.016 × 2 ..... 0.032
" (4, 3½).................... .... 0.005
" (3, 3½).................... ..... 0.000
                                         ─────
Corrected difference of temperature..... 3.177°

or 3.18°.

                                                  Calories.
Heat disengaged....3.18°.   3.18 × 2.681 =   8.5256
Iron wire...........0.025. 0.025 × 1.6 = 0.04
Nitric acid ....0.15.   0.15 × 0.23 = 0.0345
                                           ─────
                                                   0.0745
                                                  ─────
For one gram...................... 8.4511

or 8451.1 for 1 kilogram, equivalent to 15212 B. T. U.

EXAMINATION OF A GAS.

Illuminating gas was examined under the following conditions:*

Barometric pressure........... 761 mm. (29.6 inches).
Tension of aqueous vapor. ..... 8 " (0.314 inch).
Temperature of laboratory..... 18.5° C. (65.3° F.).
Volume of bomb.............. 654† cc. (39.9 cubic inches).
   "   "   "   dry at 0° and 760 mm.
                              606 cc. (37 cubic inches).

The capsule was left in its usual place in the bomb to prevent specks of iron oxide from dropping on the enamel and injuring it.

---

* See Kroeker's calorimeter on page 73.
† Exactly 653.9 cubic centimetres.

Preliminary Period.	Combustion.	After Combustion.	Remarks.
minutes. degrees.	minutes. degrees.	minutes. degrees.	
0 18.80	4½ 19.50	8 20.07	Pressure of oxygen
1 18.80	5 20.00	9 20.06	5 atmospheres
2 18.80	6 20.08	10 20.06	grams.
3 18.80	7 20.81	11 20.055	Nitric acid.... 0.06
4 . 18.80	maximum	12 20.05	Iron wire..... 0.025
$a_0 = 0.00$		$a_t = \dfrac{20.08 - 20.05}{5} = 0.006°$	

Gross difference of temperature, Δ.................... 1.28°
Correction as usual, a............................... 0 015

Difference, $\Delta + a$............................. 1.295°

                                                        Calories.   Calories.
Quantity of heat observed, 1.295°......... 1.295 × 2.681 =   3.47189
Heat of HNO$_3$ formation.................. 0.06 × 0.23   = 0.0138
Heat of iron-wire combustion............. 0.025 × 1.6   = 0.04
                                                          ———
                                                          0.0538

Heat of combustion of 606 cc. at 0 and 760 mm.............. 3.41809
or per cubic metre at 760 mm. 5640, or 633.6 B. T. U. per cubic foot.

COMBUSTION USING AN AUXILIARY SUBSTANCE.

Sometimes an unconsumed residue is left while determining the heat of combustion of some difficultly burning substances, diamond or graphite for instance. In this case a combustible auxiliary is used to obtain complete burning of the sample. The most convenient to use is naphthalin ($C_{10}H_8$), the heat of combustion of which is exactly known, 9692 calories.

Take petroleum coke, which is nearly allied to graphite. It is mixed with a little naphthalin which has been previously melted at a low heat and then cooled. After cooling the weight of the naphthalin is taken.

The coke analyzed as follows:

Carbon.......................	97.855 per cent.
Hydrogen.....................	0.489 " "
Oxygen.......................	1.196 " "
Nitrogen......................	0.260 " "
Ash..........................	0.200 " "
	100.000 " "

The data obtained are as follows:

Preliminary Period.		Combustion.		After Combustion.		Remarks.	
minutes.	degrees.	minutes.	degrees.	minutes.	degrees.		grams.
0	22.05	5½	22.60	10	25.12	Napthalin............	0.034
1	22.05	6	24.20	14	25.05	Iron wire	0.025
5	22.04	7	25.02			Nitric acid..........	0.080
		8	25.13			Water of calorimeter.	2200.
		9	25.14			Equivalent in water..	481.
$a_0 = -0.002$		maximum		$a_t = 0.015$			

$$
\begin{aligned}
\text{Difference of temperature} &\ldots\ldots\ldots\ldots\ 25.14 - 22.04 = 3.100° \\
\text{Correction for minutes } (9, 8), (8, 7), (7, 6) &\ldots\ 0.015 \times 3 = 0.045 \\
\text{`` `` ½ minute } (5\tfrac{1}{2}, 6) &\ldots\ldots\ldots\ldots\ = 0.005 \\
\text{`` `` ¼ `` } (5, 5\tfrac{1}{2}) &\ldots\ldots\ldots\ldots\ = 0.001 \\
\hline
\text{Corrected temperature difference} &\ldots\ldots\ldots\ldots\ 3.151°
\end{aligned}
$$

Then,

Total heat developed 3.15°......... $3.15 \times 2.681 =$ 8.4451

From this subtract

Heat due to naphthalin............ $0.034 \times 9692 = 0.3295$
     "   "   " iron wire............. $0.025 \times 1.6 = 0.04$
     "   "   " HNO_3................ $0.08 \times 0.23 = 0.0184$
                                                             0.3879

Heat developed by the combustion of the coke........... 8.0572
or 8057.2 per kilogram, or 14503 B. T. U.

When the combustible tested contains hydrogen, it must be remembered that, while the gas in the bomb is dry at the beginning, it is saturated at the close of the experiment. In reality, the latent heat of vaporization of the small quantity of water necessary to be added is inconsiderable. The mean of several tests was 5 in 8500 calories observed, or only $\frac{1}{1700}$. Still, when we test gases, which cause less marked difference in temperature than solids or liquids, we must allow for this heat of vaporization to be exact.

It may be asked if any allowance will be made for the heat of the electric current at the moment of kindling. The heat developed by a current with intensity I and electromotive force E is

$$C = \frac{EI}{4.17}t,$$

t being reckoned in seconds. If t was appreciable, this should be considered at least in exact determinations. But, actually, t is very small; the contact is hardly established before the iron is burnt and the contact broken.*

Mahler cites two successive tests made on the same coal with his bomb and with the bomb of the College of France, as furnishing proof of the accuracy of his method.

The following results were obtained:

	Scheurer-Kestner at the College of France.	Mahler.
Coal (pure) from Bascoup, Belgium....	8828	8813

The calculations may be rendered simpler and the observation more rapid, still being exact enough for industrial uses.

Take the equation

$$Q_0 = (\varDelta + a)(P + P') - (0.23p + 1.6p'), \quad . \quad . \quad (1)$$

arranging the terms in order of the corrections

$$Q_0 = \varDelta(P + P') + a(P + P) - (0.23p + 1.6p'). \quad (2)$$

It is clear that the calculation of the calorimetric operation

---

* In exact researches this heat can be easily determined if wished. It will be sufficient to measure the electromotive force in volts. Then put an amperemeter in the line which connects the bomb and kindle the combustible as usual. The displacement of the needle shows the intensity of the current under the conditions of the test, and also the time during which the current was closed. The formula $\frac{EI}{4.17}t$ will give the quantity of heat sought.

reduces to the determination of a maximum and to one multiplication if we have

$$a(P + P') = 0.23p + 1.6p'. \quad . \quad . \quad . \quad (3)$$

Now from the tests made we readily see that whatever value a may take, it increases with the quantity of heat generated in the bomb; it is a little greater when the external air is warmer than when it is cooler—a fact which may be attributed to the influence of evaporation on the cooling of the bath.*

On the other hand, the nitric acid appears to increase with the quantity of heat generated, and tends to offset the correction from a. In short, p' is, within certain limits, at the control of the observer, same as P'. We consider it then possible to arrange once for all so as to have the expression (3) sufficiently close for industrial purposes.

This can be done with Mahler's apparatus. Thus for oil of colza the multiplication $\Delta(P + P')$ gave 9625 calories, which is within $\frac{1}{3000}$ of the final number obtained after all corrections; with the Nixon's coal we found that $\Delta(P + P') = 8418$ calories, which differed $\frac{1}{250}$ from the correct number; with coal-gas the product $2681 \times 1.28 = 3432$ calories, while the corrected result was 3418, or $\frac{1}{240}$ difference.

ATWATER'S CALORIMETER.

Prof. Atwater has considerably modified the bomb, so that it seems to have some advantages for easy working. Fig. 25 gives a sectional view of it in the calorimeter. The steel used is the same as that used in the Hotchkiss guns,

---

* The rapidity of cooling in the apparatus employed by Mahler was, according to experiments, between 15° and 20° C.

$$\frac{d\theta}{dt} = 0.005(T - T_0),$$

T_0 being the temperature at which cooling ceases.

and having an unusually high tenacity, seems admirably fitted for the purpose. *A* represents the bomb, *C* the screw-cap, *B* the cover, which is placed on the bomb cylinder and held down by the screw-cap. "The cover is provided with a neck into which fits a cylindrical screw *E*, holding another screw *D*. *F* On the side of the neck is an aperture *G*, between the lower end of *D* and the shoulder. In *D* is a washer of lead, on which the lower edge of *E* fits. By opening or closing the screw *F* the narrow passage from *z* is opened or closed. The opening is used for admitting oxygen at a high pressure through a narrow passage to charge the bomb. In *B* is an aperture through which passes the platinum wire *H*, which is separated from the metal of the cover by insulating material. Hard vulcanized rubber serves very well for this purpose. Fastened to the lower side of the cover is another platinum rod, *I*, between which and *H* an electrical connection is made with a very fine iron wire. A screw-ring holds the small platinum capsule, in which the substance to be burned is placed. At *KK* are ball-bearings of hard steel to avoid friction in screwing the cap down."

FIG. 25.—ATWATER BOMB.

"The large cylinders *N* and *O* are made of indurated fibre, and covered with plates of vulcanized rubber. A stirrer serves for equalizing the temperature of the different portions of water after the combustion is completed." *

The thermometer used is by Fuest of Berlin, graduated to $\frac{1}{100}$ degree, and can be read with a magnifying-glass to $\frac{1}{1000}$ degree.

---

* Prof. W. O. Atwater, in Bulletin No. 21, U. S. Dept. of Agriculture, 1895, pages 124 and 126.

The apparatus has been used with success in making the very numerous determinations made by Atwater on the heats of combustion of food-products and other allied organic substances.

KROEKER'S CALORIMETER.

Kroeker has recently modified the bomb, making two inlet channels instead of one. By this means he has a current of oxygen gas passing in at one opening and waste gases passing out at the other. It can thus be used for the same purpose that a Junker calorimeter is used, and it is claimed with just as satisfactory results.

The cylinder (Fig. 26) is bored out of a piece of Martin steel, and has a closely-fitting screw-plug for cover, the depth of the screw joint being 25 mm. The walls of the cylinder are 10 mm. thick; external diameter, 72 mm.; internal diameter, 52 mm.; height, 120 mm.; contents, 200 cc. It has four small legs on the under side, which support it and keep it entirely surrounded by the water of the bath. The entire inside surface is enameled, or preferably platinized. The fuel, in the form of compressed cylinders weighing one gram, is put into the carrier, ignited as usual, and the combustion gases collected and examined.

He also has a method of heating the calorimeter bomb in an oil-bath so as to expel all the water of combustion and hydration. He thus obtains data for corrections due to the usual method of determining the water, i.e., considering the water as condensed.

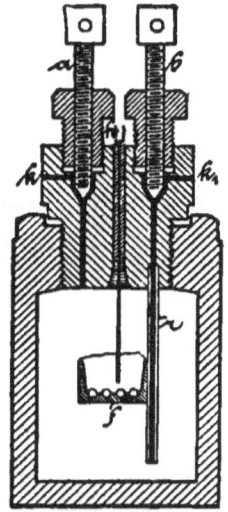

FIG. 26.—KROEKER CALORIMETER.

WALTHER-HEMPEL BOMB.

Two modifications of the Berthelot bomb are known under this name. The larger one does not differ in enough points to make a special mention of it necessary; but the smaller one, the one intended for use in analysis, is worthy of description.

It consists of a small cylinder of 33 cc. capacity (Fig. 27), bored out of white cast iron and enameled inside. The walls are 2 millimetres thick, and it is strong enough to resist eight times the pressure generally used. The cover is fastened on by means of a screw-clamp, and through it passes the slanting opening a, having the electric wire-carrier insulated by a caoutchouc sheath. To the wire at the end of this sheath is attached a platinum wire for kindling the combustible. On the opposite side of the cover is the oxygen tube d. The platinum wire c is attached to the under side of the cover, and supports the combustible-carrier and its little fire-clay cylinder e.

FIG. 27.
WALTHER-HEMPEL BOMB.

The fuel is made into small cylinders by compression, put into the fire-clay cylinder, and ignited by the electric spark. The products of combustion are collected and weighed or measured: the water partly in the bomb and partly by means of a calcium chloride tube; the nitric and sulphuric acids are determined by titration with $\frac{1}{100}$ normal alkali, and afterwards separated if deemed necessary. It is claimed to be capable of use the same as a large one. A full description of it is given in the *Berliner Bericht* for January, 1897.

CHAPTER VII.

SOLID FUELS.

COAL.

AMONG the first careful tests ever made, to determine the heat value of different kinds of coal, are those made in 1843 and 1844 by Prof. W. R. Johnson for the U. S. Navy. He analyzed and tested all the kinds obtained from the United States and England, which were then in use by the navy. At the time they were made the calorimetric determinations were not considered as of the importance they are now, and his tests were limited to determining the evaporative power of the coals. Mr. W. Kent reviewed them in the *Engineering and Mining Journal*, 1892, and showed that up to the time of the experiments nothing comparable with them had been attempted, and that in many respects they compare favorably with work done to-day.

In 1857 Morin and Tresca made numerous determinations of the calorific power of coal and wood, and in 1853 they published a work on "Fuels and their Calorific Power," in which they make many recommendations for more accurate work. They wrote: "It would be extremely important if experiments with the calorimeter could be made on most of the fuels, by methods similar to those used by Favre and Silbermann."

In 1868 such experiments were made by Scheurer-Kestner, and continued by him later with the aid of Meunier-Dollfus. They based their calculations on pure coal, i.e., with moisture and ash deducted. This method, which has been followed by many others, seems very logical, as it facilitates comparison of different fuels by reducing them to the same basis. Enormous errors due to comparison of values not comparable are thus obviated. Coal having 5 per cent impurity has been compared with coal having only 1 per cent, no account being made for the difference, and of course very erroneous and misleading deductions obtained.

It is a simple task for the engineer or the workman even, to determine approximately the proportions of moisture and ash as given on the grate. Knowing these proportions and the heat of combustion of the pure coal, they can render a statement of the practical working. If, on the contrary, the experimenter is limited in such way that he neglects the composition of the coal, it is impossible to make a conjecture as to its intrinsic or comparative value; still less can he judge of it as a steam generator.

In 1879 Bunte made some experiments at Munich, using a special apparatus devised by him for the occasion, which was part calorimeter and part boiler. The tests were published in Dingler's *Polytechnisches Journal*. Some of the results are included in the tables of this book.

Since then numerous tests have been made on nearly all the known coals. A collection of all available ones from which the desired data could be obtained will be found farther on.

The question as to the actual evaporative effect of each coal can be settled only by actual tests made on the boiler intended for use, as the same coal will give slightly different results with different kinds of boilers; also, and in a more marked degree, with different methods of firing and handling. The results in the tables cannot be taken, then, as absolute for all boilers under all circumstances, but they can be depended on for comparison of the different fuels with the same boiler and under proper conditions.

The manner in which a coal acts under heat in a closed

vessel is a most important indication, taken in connection with its elementary composition. Gruner gave his opinion that the real value of a coal could be determined better from its proximate than from the ultimate composition. Speaking of the Loire coal, he says:

"The proximate analysis, which consists in distilling coal in a retort and incinerating the residue, allows direct valuation of the agglomerating power as well as the nature and proportion of the ash. Further, it is easy to show, especially with the aid of the work of Scheurer-Kestner and Meunier-Dollfus, that the calorific power varies with the proportion of fixed carbon left by distillation. This is true at least for all coal properly so called, but not always true for anthracite and lignite." *

Gruner formed the following table based on the quantity and nature of the coke furnished and the calorific power. He held, from the results of S.-K. and M.-D., that if the heat value of a coal increases with the proportion of fixed carbon

Classes or Types of Coal properly so called.	Per Cent Coke to Pure Coal.	Per Cent of Volatile Matter in Pure Coal.	Nature and Appearance of Coke.	Calorific Power, Actual. Calories.	Industrial Calorific Power. Water at 0° Vaporized at 112° per Kilo of Pure Coal Burnt, in Kilograms.
1. Dry coals with long flame,	55 to 66	45 to 40	Powdery or slightly coked.	8000 to 8500	6.7 to 7.5
2. Fat coals with long flame (gas coals),	60 to 68	40 to 32	Completely agglomerated, often-er caked, but porous.	8500 to 8800	7.6 to 8.3
3. Fat coals, properly so called ("blacksmith" coals),	68 to 74	32 to 36	Caked and more or less puffy.	8800 to 9300	8.4 to 9.2
4. Fat coals with short flame (coking coals),	74 to 82	26 to 18	Coked, compact.	9300 to 9600	9.2 to 10
5. Lean coals or anthracite,	82 to 90	18 to 10	Slightly coked, oftener powdery.	9200 to 9500	9.0 to 9.5

* Annales des Mines, 1878, vol. IV.

or coke formed, this increase is produced gradually by cutting off the lean coals and dividing the fat coals into three classes —gas, forge, and coking.

Bearing on the advisability of having proximate analyses, as well as ultimate analyses of coal, is the question recently brought up by Mr. Kent, regarding the ratio of hydrogen and carbon in coal. In discussing the results of Lord and Haas' determinations of Ohio and Pennsylvania coals, he thought he had discovered the ratio, that the fixed carbon is nearly equal to the total carbon minus five times the available hydrogen in bituminous coals, and minus three times the hydrogen in semi-bituminous ones. He gave a table showing results which support the hypothesis.

LIGNITE.

From an industrial standpoint lignite is of considerable importance. It occurs in most countries, and is used in a great many for domestic and manufacturing purposes.

As a fuel it is inferior to coal, being less distantly removed from woody fibre, and hence contains more hydrogen and, usually, considerable water. Most of the latter, however, dries out on exposure to the air. In some cases as much as 40 or 50 per cent of water is found in the freshly mined lignite, of which at times 20 per cent remains when air-dried. This greatly affects its value as fuel; still it is used in many of the Western States, and also in Europe. In some European localities, when thoroughly dried and compressed into blocks, especially in Italy and Austria, it is used as fuel for producing gas and for evaporating, with good results. In Austria it is burnt without any preparation, except drying in the air for heating saltpans.

The amount of ash varies exceedingly, being in some cases as low as 0.9 per cent, and in others as high as 58 per cent. It even varies in the same locality and in the same bed. In burning lignite there is considerable loss in the waste gases on account of the large quantity of air introduced, and also from the moisture carried off from the fuel.

Brix published the following results with dried lignite:

	Water Evaporated.	Per cent Ash.
Lignite of Aussig, Bohemia	5.8 pounds	15.0
" " Perleberg. "	5.6 "	6.0
" " Goldfuchs n. Frankfort	5.5 "	9.1
" " Rauen	5.4 "	6.3

Bunte used two kinds of lignite in boiler-tests, and gives the following results:

	Neusattel.	Chodan.
Calories in steam	42.8	49.2
" " gases	19.6	21.0
" " aqueous vapor	9.2	8.7
" " ash	9.0	6.1
" unaccounted for	19.4	15.0

The grate used was a step grate (Treppen-Rost).

The lignite used on the railways in Italy contained 15 per cent of water, and gave a yield of heat equal to one half its weight of coal.

Analogous to the lignites are certain shales or fossils carrying bitumen. They are sometimes termed *boghead cannel*, *bituminous schist*, etc. They are distilled in some localities for oil, but are not much used as fuel.

Bunte determined the heat of combustion of a sample from Australia, and analyzed one from Scotland.

	Carbon.	Hydrogen.	O + N.	Calories.
Boghead shale, Australia.	83.17	10.04	6.79	9134
Scotch Boghead	81.54	11.62	6.84	

Scotch Boghead generally contains 18 to 24 per cent of ash. From its analysis as above, its heat of combustion should be near that of the other one given.

PEAT.

Peat is formed by the agglomeration of vegetable *débris*, and retains a large amount of water, which will not separate without heat. Its composition varies but little from that of wood, the principal difference being less oxygen and more carbon.

The composition may be represented by—

Carbon...............................	60
Hydrogen	6
Oxygen and nitrogen...............	34
	100

The heat of combustion is lower than that of coal or lignite, as might be expected. The quantity of hydrogen exceeds that necessary to form water with the oxygen.

It is usually dried before using, and when dry becomes quite porous. It carries, however, in this state some 10 to 15 per cent of water, which can be expelled only by artificial means. Large quantities of it are converted into charcoal in special kilns, and, where the large amount of ash is no objection, it makes a good fuel. It cannot be used for metallurgical purposes on account of its friability. From 30 to 40 per cent of its weight is left in the charcoal as carbon, but at the same time the ash increases to 15 to 25 per cent, and even more. This consists principally of phosphates and sulphates, with very little carbonates; hence it is not as apt to clinker as other fuel ashes.

Brix obtained with peat an evaporative power of 5.11 pounds of water. The peat used was from Flatow, and contained 10.7 per cent of ash. Another, from Buchfeld-Neulangen, contained 1.2 per cent of ash, and gave 5.12 pounds evaporated. Noury, using a special grate, obtained from the Alsace peats 4 to 5 pounds evaporation (ashes deducted).

Bunte analyzed the gases produced by the combustion of peat on the hearth of a salt-pan, and found, carbonic acid 13, oxygen 6.4, nitrogen 80.6.

Karsten says that $2\frac{1}{2}$ pounds of peat are equal to one of coal. In some experiments made at St. Petersburg a firegrate of 32 square feet and 696 square feet of boiler heating surface was used. The peat was compact, hand-moulded into 4-inch balls, and dried till moisture did not exceed 14 per cent. 4.26 pounds of coal were evaporated for 1 of peat.

Crookes and Rohrig, in their "Metallurgy," say: "One pound of dry turf will evaporate 6 pounds of water. Now in 1 pound of turf, as usually found, there are $\frac{3}{4}$ pound of dry turf and $\frac{1}{4}$ pound of water. The $\frac{3}{4}$ pound can evaporate $4\frac{1}{2}$ pounds of water; but out of this it must first evaporate the $\frac{1}{4}$ pound of water contained in its mass, and hence the water boiled away by such turf reduces to $4\frac{1}{4}$ pounds. The yield is here reduced 30 per cent, a proportion which makes all the difference between a good fuel and one almost unfit for use. When turf is dried in the air under cover it still retains $\frac{1}{10}$ of its weight of water, which reduces its calorific power 12 per cent; 1 pound of such turf evaporates $5\frac{1}{8}$ pounds of water."

COKE.

Coke usually met with is from three sources: from gas-coal, and made in gas-retorts; from gas or ordinary bituminous coal, and made in special ovens; from petroleum, and made by carrying the distillation of the residuum to a red heat.

Coke from gas-works is usually softer and more porous than the other kinds, burns more readily, but does not give as intense a heat. It has been used considerably for domestic heating, and in factories where a high heat is not needed but where a smokeless fuel is desirable. The oven coke is usually in large columnar masses of a close texture and quite hard. It has a dead gray-black color and is not susceptible of polish. It is principally used in furnaces requiring a blast, although limited quantities of it have been used in domestic heating, for which purpose it must be broken up much finer than its usual size. Petroleum coke is generally in large irregular lumps, perforated with cavities of greater or less size, the interior of which is usually quite smooth and shining. Its color is blacker than that of gas or oven coke, and its hardness intermediate. It is used principally for making electric carbons, although considerable quantities are used for fuel.

With the exception of gas-coke very little use is made of this fuel for steaming, the fire being too intense locally, and hence very apt to burn out the boiler directly over it. In all cases plenty of air is needed to keep up the combustion, which is also a drawback for steaming purposes. For metallurgical furnaces it is different. Here it is almost the ideal fuel, giving an intense reducing heat at just the part of the furnace where most needed. It has been used in iron furnaces for years, and is still the favorite fuel. It is superior to anthracite, as it has no tendency to splinter and crack with the heat, and bears its burden very well. Of course this does not apply to ordinary gas-coke, which crushes easily.

Coke is essentially carbon, and the mineral portions of the coal from which it is made. It contains small quantities of hydrogen and nitrogen, as may be seen from the tables. The percentage of these, however, is very low, so that the calculated and observed heat-units are usually within the limits of error, as is shown in the following table:

Name.	C.	H.	N.	Loss.	Calories observed.	Calories calculated.	Authority.
Saarbruck.....	98.04	0.73	....	1.23	8200	8229	Bunte
Petroleum coke	98.05	0.50	0.25	1.20	8057	8151	Mahler
Graphite	98.98	0.02	....	....	7901	8054	Berthelot

WOOD CHARCOAL.

Wood charcoal always contains quantities of hydrocarbons which have resisted the action of heat. That called forest charcoal, made by burning in heaps, is the most charged with them; that obtained from distillation of wood in retorts contains less.

The heat of combustion is very variable. According to Berthier* commercial wood charcoal contains 10 per cent of volatile matters and 2 per cent of ash (carbon 80 to 90, hydrogen 1.5–4).

Pure wood charcoal was first tested calorimetrically by Favre and Silbermann, and since then by several experimenters. To obtain it pure it was calcined strongly and treated with chlorine to remove all traces of hydrogen. In this state wood-charcoal produces under constant pressure 8080 calories, F. & S., or 8100 S.-K. & M.-D.; with constant volume Berthelot and Petit obtained 8137 calories.

Several years ago Berthier pointed out that half-burnt charcoal, *charbon roux* or *Rothkohle*, was superior in combustible content to that perfectly burnt. Sauvage has confirmed this, and gives the following results:

100 lbs. of wood charred for......	3 hours.	4 hours.	5 hours.	5½ hours.	6½ hours.	Mound Charcoal.
Weighed.........	65.4 lbs.	53.0 lbs.	47.0 lbs.	41.5 lbs.	39.1 lbs.	17.2 lbs.
100 cu. ft. measured	86 cu. ft.	76 cu. ft.	58 cu. ft.	55 cu. ft.	52 cu. ft.	33 cu. ft.

and

1 cubic foot wood contained of combustible matter					908 parts.	
1 " " 3 hours' heating "	"	"	"	883	"
1 " " 4 " "	"	"	"	904	"
1 " " 5 " "	"	"	"	1133	"
1 " " 5½ " "	"	"	"	1091	"
1 " " 6½ " "	"	"	"	1136	"
1 " " charcoal "	"	"	"	1069	"

* Traité des essais par la voie sèche, vol. I, p. 286.

So that the amount of combustible matter does not increase after 5 hours' heating, and a continuance of the heat diminishes it.

The principal use of charcoal is in iron furnaces, where it has been used for years, and produces the highest grades of iron, being free from sulphur and phosphorus. A small amount is used in private dwellings and hotels for heating and cooking. For boiler heating it has been used only experimentally.

Scheurer-Kestner and Meunier-Dollfus experimented with it in boiler-heating and found very little combustible gas in the products. Beech charcoal was used, and an evaporative effect of 7.62 pounds of water was obtained. The waste gases contained:

Carbonic acid	11.16 per cent.
Carbonic oxide	0.37 "
Oxygen	8.72 "
Nitrogen	79.75 "
	100.00

Brix, using wood and peat charcoal, obtained the following results:

Wood charcoal	7.55 pounds evaporated.
Peat charcoal	6.85 " "

Schwackhöfer burnt charcoal from hard and soft wood in his calorimeter and obtained (constant volume) 7140 calories for the soft charcoal and 7071 calories for the hard. The charcoal in both cases was the ordinary unpurified charcoal as sold.

WOOD.

Wood consists of a compact tissue more or less hard, formed of cellulose and a so-called incrusting substance.

Wood contains, besides, small quantities of mineral matter and hygroscopic water varying from 15 to 30 per cent, according to dryness. Air-dried, it contains about 15 per cent of water, which it gives up easily on exposure to a heat of 100° C.

The composition of wood may be represented by the following:

	Carbon.	Hydrogen.	Oxygen.	Ash.	Water.
Wood dried at 100°	49.5	6.0	43.5	1.0	0.0
" " in the air	29.6	4.8	34.8	0.8	29.0

Regarding wood from its ultimate composition, we may consider it as a hydrate of carbon, that is, as carbon united to water, the proportion of hydrogen and oxygen being nearly the same as in water. But regarded from its proximate composition, it is entirely different. What has been said of soft coal can be repeated for wood; that, those having a similar ultimate composition behave differently in distillation in a closed retort and produce very different proportions of carbon (as charcoal); hydrocarbons, liquid or gaseous; acid products, resin, and tar. It was supposed that the heat of combustion differed also, and this has been verified by experiments.

Berthelot and Vielle determined the heat of combustion of cellulose, and found 680 calories for the molecular weight of wood, or about 4200 calories per kilogram.

Hard wood gives less heat than soft wood. According to Gottlieb's experiments, pine-wood has a heat value of 5000 calories, while oak gave only 4620 calories. Mahler's experiments confirm a difference in favor of pine, but in less proportion.

Two determinations made by Mahler are (cinders and water deducted):

	Fir.	Oak.
Carbon	51.08	50.43
Hydrogen	6.12	5.88
Oxygen with trace of nitrogen	42.90	43.69
	100.00	100.00
Heat of combustion	4828	4689

Gottlieb obtained the following numbers, using a calorimeter of constant pressure, in which he burnt 2 grams of wood in the space of two or three minutes. The composition of the gas produced was not determined; he was satisfied that he had perfect combustion, and his figures do not appear very far from the truth. For cellulose he obtained 4155 calories.

Name.	C.	H.	N.	O.	Ash.	Calories.	B. T U.
Oak	50.16	6.02	0.09	43.36	0.37	4620	8316
Ash	49.18	6.27	0.07	43.91	0.57	4711	8480
Elm	48.99	6.20	0.06	44.25	0.50	4728	8510
Beech	49.06	6.11	0.09	44.17	0.57	4774	8591
Birch	48.88	6.06	0.10	44.67	0.29	4771	8586
Fir	50.36	5.92	0.05	43.39	0.28	5035	9063
Pine	50.31	6.20	0.04	43.08	0.37	5085	9153

Gottlieb's results are 69 calories less than Mahler's for oak and 207 more for fir.

In burning wood for steaming the fire is easily controlled; combustion is more complete; the products of combustion contain only very small quantities of unburnt gases; and the ashes are generally free from carbon. The countries using wood for this purpose are growing less in number yearly, on account of improvement in transportation and the discovery of new coal seams; petroleum oils for fuel have also become more common, especially in Russia, the United States, and Canada.

Morin and Tresca, in their tests, found that one pound of wood was equivalent to 0.368 pound of coal. Scheurer-Kestner's experiments in 1871 show results more favorable for wood. The wood used was Vosges fir, which had been piled under cover for half a year. A cubic foot weighed 19.76 lbs. It was burnt in the same boiler used in his previous experiments, with the result that 1 pound of wood evaporated 4.4 pounds of water. The ratio was 0.490, or nearly one half that of Ronchamp coal.

Brix made a number of experiments in using wood for heating, and found that dry pine gave the best results—5 pounds per pound of fuel. Elm gave 4.6 pounds; birch, 4.6; oak, 4.56; ash, 4.63; and beech, 4.47.

Wood should be dry as possible, as otherwise it has to furnish heat to vaporize, not only the water formed from its hydrogen, but also that already existing as moisture. We have seen that this loss with coal is considerable, it is still greater with wood. Suppose the wood to be ordinary air-dried, containing 20 per cent of water. If this wood, when perfectly dry, could evaporate 5 pounds of water, it now has only $\frac{4}{5}$ of that power, or power to evaporate 4 pounds; but it already carries $\frac{1}{5}$ of its weight of water, which must be vaporized. Hence the available power is 4 pounds less $\frac{1}{5}$ pound = $3\frac{4}{5}$ pounds, or 76 per cent of its dry value. Hence the economy of using only dried, and even artificially dried, wood.

CHAPTER VIII.

LIQUID FUELS.

SHALE-OILS.—PETROLEUM.

THE mineral oils comprehend the liquid hydrocarbons extracted from bituminous schist or coal and its congeners by distillation, as well as the oils which exist already formed in the earth, and called by the special name of *petroleum*.

While the former are seldom employed in heating, petroleum has become an important fuel in the countries which produce it. Its special qualities, light weight, and low price per calorie compared with other fuels insure a great future. The knowledge of its heat of combustion has become, then, of considerable interest.

Its ultimate percentage composition varies within rather close limits, yet it is of a very complex proximate composition. The industry of refining crude petroleum extracts from it some 50 per cent of refined oil for use in lamps, and having a density of 45° to 47° Beaumé, boiling-point 170° C. (328° F.); 10 per cent of naphtha with a lower density and boiling-point; and 20 per cent of paraffin oil of a higher density and boiling-point.

Crude petroleum contains a large number of hydrocarbons of the general formula C_nH_{2n+2}, and running from CH_4 to $C_{11}H_{44}$, with many isometric modifications. The industrial treatment modifies it profoundly. Hydrocarbons containing 95 per cent of carbon have been found in the products of distillation.*

---

* Wurtz, Dictionnaire de Chimie, Supplement.

The first calorimetric experiments were published by Ste.-Claire Deville in 1868 or 1869, using a large calorimeter especially constructed for the work. Mahler used the bomb. The liquids were burnt in the bomb under nearly the same conditions as solids, when they had no appreciable vapor tension. When they had considerable vapor tension (light oils, for instance) Berthelot placed them in a closed vessel, the bottom being platinum and the top formed by a pellicle of gun-cotton.

Heating by oil is quite recently introduced, but is already developed to a high degree in Russia and on this continent, and is gaining in other localities. The small volume occupied in comparison with its high calorific power renders it a formidable competitor with coal.

To burn petroleum, atomizers fed by steam or compressed air are used. They generally consist of a horizontal pipe under the boiler, fed with oil from an elevated reservoir placed at a presumably safe distance. The steam enters inside the oil-pipe, and, mixing with the oil, throws it into a spray and produces a flame several feet long. At the Chicago Exposition 52 tubular boilers were exhibited heated by oil, developing a power of 25000 H.P., and yielding a total evaporation of 12000 cubic feet per hour. The oil used was the heavy portion of petroleum (the lighter ones having been distilled off for illumination), and it was fed under a pressure of one-fourth atmosphere. The result was an evaporation of about 15 pounds of water per pound of oil.

In 1889 Albert Hubner ran a whole battery of boilers with oil at his works in Moscow. He used Baku Nafta, or "Mazoute," which contained carbon 86.3, hydrogen 13.6, and oxygen 0.1 per cent. The density was 0.910 to 0.914.

At Petrolea and Oil City, Canada, the heavy residuum from the stills is used as fuel under boilers and stills. The burners used are very simple, and run without producing smoke. In the United States, the Standard Oil Company has pushed the sale of fuel-oil made of Ohio crude, and large quantities of it have been used; large quantities of a special grade are also made for use in enriching water gas.

The calorific power of petroleum residuum is, according to Sainte-Claire Deville, 11460 calories (20628 B. T. U.), the evaporation at 5 pounds pressure being 15 pounds. This compared with the heat of combustion shows a useful effect of over 86 per cent, while the entire absence of smoke, unburnt gases, ashes, and irregularity in air-supply add to its advantages still more.

Some experiments made at the Hecla Engineering Works, Preston, England, and lasting two days, used a marine boiler. The first day natural draft was used, the second a Körting blower. The oil was blast-furnace oil from Sheffield, and contained:

	Per cent.
Carbon	83.54
Hydrogen	10.59
Oxygen	5.94
Sulphur	0.09
	100.16

By Thompson's calorimeter its value was 16080 B. T. U.
Equivalent to water at 212 °F .... 16.66 pounds.

The results were: First day, 14.97 lbs.; second day, 14.21 lbs.,—a yield of 89.87 and 85.25 per cent of the theoretical.

A series of tests made at South Lambeth with a Cornish boiler showed 20.8 lbs. evaporation; average of several days, 19.5 lbs. The same boiler with the best Aberdeen coal yielded 6.5 lbs.,—an advantage of 3 to 1 in favor of the oil.

The following analyses of the waste gases from boilers using oil show how perfect the combustion is, and that little if any excess of air is needed:

CO_2	14.19	18.08
CO	5.20	0.34
O	0.78	0.34
Hydrocarbons	1.30	None.
H	Not determined.	None.
N	78.53	81.24

To have the best results, the burner must be so regulated as to have a flame bordering on, but not quite, smoky. Thus sufficient and not too much air is obtained. The quantity of steam needed to atomize the oil at Moscow is 4 per cent of the water evaporated. The use of compressed air has been tried in some places with very satisfactory results: the atomizing is good, but the cost is higher, and the probable chemical effect of the steam is wanting.

Nothing but a bare mention need be made of animal and vegetable oils, as they are not used in the arts for heating purposes except, perhaps, on very exceptional occasions. The calorific power of all of them is high, as may be seen from Table I.

CHAPTER IX.

GASEOUS FUELS.

THE heat of combustion of gaseous combustibles has been determined for a great many compounds, definite and pure. That of the industrial gases has been determined by different operators and in different ways, with more or less happy results. Its determination is often one of the greatest commercial interest, since it is used in domestic heating as well as in industrial appliances, where it is necessary to obtain definite, regular working. It serves also to furnish motive power to gas-engines, in which the heat of combustion is not without importance. Finally, it is well to know the heat produced in air or water-gas apparatus, if we wish to reach the best condition for their production and use.

For heating steam-boilers gas has given good results and a very high evaporative effect. It is easily regulated, and thus any required heat can be produced by simply turning a valve. No smoke is generated, no soot or deposit of any kind produced in the flues, and no ashes to take out of the ash-pit. The fireplace needs repairing but seldom, and the boiler is heated evenly and regularly, there being no danger of burning out in strongly heated spots, as no such spots exist.

In metallurgical furnaces, gas possesses a decided advantage in its long, clean, easily managed, intense flame, and this advantage has been long recognized. A flame of 25 feet or more in length is easily produced, and it is practically uniform for its whole extent. Part of the heat usually lost up the chimney can be utilized to heat the air-supply, and no more is supplied than just enough for perfect combustion.

Using gas as fuel enables the metallurgist to use poor

grades of coal, and all variations in quality may be eliminated, a uniform product being had by storing the gas in a holder, or by making proper arrangement of different generators so that an average will be obtained. In several cases where hand-fed coal fires have been tried against fires burning gas from the same coal, better results have been obtained, due to the possibility of more closely adjusted regulation. The tests made at Brieg may be cited. Here each boiler had 141.25 square feet of heating-surface and steam-pressure 6 to 7 atmospheres.

No. 1 boiler was hand-fired; No. 2 was gas-fired. The evaporation in pounds per pound of fuel was:

No. 1......	8.34	8.74	8.28	4.02	2.569	2.764
No. 2......	9.86	9.73	10.07	5.44	3.251	3.158
Increase...	18%	12%	20%	35%	25%	14%

HEAT OF COMBUSTION OF GASES FROM ANALYSIS.

When the chemical composition of a gas is known exactly, its heat of combustion can be correctly calculated; but in absence of a correct analysis, the calorimeter must be used.

Knowing the proximate composition of a combustible gas, that is, the proportion of chemically defined components as well as their heats of combustion, it is sufficient to add the numbers obtained for each constituent gas. Take, for example, the analysis of illuminating gas of Manchester as given by Bunsen:

Hydrogen.........................	45.58
Marsh gas (CH_4)...............	34.90
Carbonic oxide..................	6.64
Ethylene (C_2H_4)...............	4.08
Butylene (C_4H_8)...............	2.38
Sulphydric acid.................	0.29
Nitrogen	2.46
Carbonic acid...................	3.67
	100.00

The calculation is as follows:

Components.	No. of Litres per Cubic Metre.	Weight per Cubic Metre at 0° and 70° mm. Grams.	Heat of Combustion per Cubic Metre.	Calculated Calories.
Hydrogen...............	455.8	89.61	3066	1395
Marsh gas, CH_4.........	369	715.58	9340	3169
Olefiant gas, C_2H_4......	40.8	1251.94	14980	611
Butylene, C_4H_8	23.8	2503.88	29042	690
Carbonic oxide	66.4	1251.50	3057	201
Sulphydric acid, H_2S...	2.9	2551.99	11400	33
Total calories per cubic metre........................				6099

City of Manchester gas, as analyzed by Bunsen, gives, then, with complete combustion, 6099 calories per cubic metre (685 B. T. U. per cubic foot).

If, however, only the actual ultimate composition of the gas is known or the total percentage of carbon, hydrogen, oxygen and nitrogen, then the calculated result will differ from the experimental one. This is because the heat units of the elements added together do not make those of the compound, as the heat of combination of the different constituent gases is not allowed for. If this factor is known, then it can be used as a correction and the correct heat determined.

This heat of combination of the elements to form the component gases will be seen in comparing the calculated and the actual heat of combustion of the following gases:

Gases.	Formulæ.	Carbon.	Hydrogen.	Calculated Heat.	Actual Heat.	Difference.
Marsh gas..........	CH_4	75.	25.	14685	13343	+1342
Olefiant gas.........	C_2H_4	85.7	14.3	11859	12182	− 323
Acetylene..........	C_2H_2	92.3	7.7	10114	12142	−2028
Benzene	C_6H_6	92.3	7.7	10114	12410	−2296

It will also be seen, that although two gases may have the same percentage composition of the elements, yet the heat of combustion may be different owing to the action of the various physical forces at work in molecular condensation, etc.

COAL GAS.

The heat of combustion of illuminating gas obtained from the distillation of coal in closed retorts is very variable. It depends not only on the nature of the fuel, but also on the rapidity of the distillation and the heat by which it is accomplished. The heat of combustion varies from 5200 to 6300 calories per cubic metre. It cannot be represented by any average number.

According to Witz, at the same gas-works and with the same fuel, yields may occur from 4719 to 5425 calories. According to Bueb-Dessau, the illuminating gas of the same city during the same day will sometimes vary 20 per cent. Dr. Birchmore reports the same result from his examinations of the gas of Brooklyn, N. Y.

We are not certain that the composition assigned to coal gas by analysis corresponds always to the gas as obtained by distillation; in Europe, especially, a portion of the heavy hydrocarbons is taken out for sale separately, and the deficiency supplied by cheaper oils.

From several experiments which he made, Bueb-Dessau* thought that the heat of combustion of illuminating gas was directly proportional to the candle power; but in addition to this being opposed to the theory of heat, the experiments of Aguitton show the contrary. He concluded from his determinations that each illuminating gas of different candle power has a definite heat of combustion which corresponds to the intensity of the light. His experiments were carried on with more than a hundred samples, rich and poor, the former kind from cannel coal, the latter from the end of the run carried to an extreme. He represents by the following formula the

---

* Bueb-Dessau cites the following among others:

	Candle-power.	Heat-value.
Gas of Dessau	14.	4400 calories
Gas of Bremen	21.9	5977
Gas from cannel coal	26.0	6559

relation between candle power and heat of combustion of a gas:

$$c = i \times 352.6 + 2280,$$

in which c represents the heat of combustion and i the candle power. The formula seems to be applicable only between limits at which it has been verified—from 5 to 15 candles. Aguitton's determinations were made with the calorimetric bomb.

The following table gives a *résumé* of his observations:

Candle Power.	Heat of Combustion per Cubic Metre.
5	4043
6	4395
7	4748
8	5101
9	5453
10	5806
11	6158
12	6511
13	6864
14	7216
15	7569

$$\frac{7569 - 4043}{10} = 352.6, \text{ coefficient adopted.}$$

The three samples of illuminating gas, analyzed and burnt in the bomb by Mahler and given in the table below, call for the following observations: Gas from Niddrie cannel coal, the most calorific per cubic metre is the least calorific per kilogram, because the density is greater than that of the other two. The richest in hydrogen by volume (Lavillette) is the poorest in calorific power per cubic metre, while the poorest in hydrogen by weight is the richest in calories per cubic metre. These are due to the low density of hydrogen, which is less calorific by volume than the other hydrocarbons occurring in illuminating gas.

Name.	Density.	Analysis by Weight.					Heat of Combustion	
		Carbon in Hydrocarbons.	Hydrogen.	Carbonic Oxide.	Carbonic Acid.	Oxygen, Nitrogen, and Sulphur.	Per Cubic Metre at 0° and 760 m.	Per Kilogram.
Niddrie cannel..	0.6367	43.33	13.50	16.84	9.26	14.96	6365	7735
Commentry coal.	0.4046	43.74	21.46	24.96	7.08	5.75	5824	11100
Lavillette gas...	0.4033	42.25	21.34	21.23	6.83	8.33	5602	10764

A cubic metre of hydrogen develops 3091 calories in burning; a cubic metre of marsh gas develops 10038 calories; a cubic metre of olefiant gas, 15250 calories.

GAS OF GASOGENES.

The gasogenes, instead of transforming the fuel into carbonic acid and water in a single combustion, produce this change in two distinct burnings, the first being to make a combustible gas and the second to burn this gas with air.

In the first furnace, the coal, for example, is burnt in such a manner by feeding with an insufficient supply of air that a gaseous mixture is produced, containing principally carbonic oxide, besides nitrogen from the air. As the combustion has been well or poorly managed, it contains a less or greater quantity of carbonic acid, the production of which is avoided as much as possible. This is done by giving to the fuel only just enough air to form carbonic oxide, and not enough to form carbonic acid, even partially, and by making the bed of fuel quite deep.

The heat produced by this combustion is not used, and consequently an important part of the calories of the coal is lost. Gasogene gas is then lower in calories, and inferior to coal gas, as commonly made by distillation.

One kilogram of carbon burnt to carbonic oxide disengages 2489 calories, while 1 kilogram of carbon burnt to carbonic acid generates 8137 calories. There is lost, then, in burning carbon to carbonic oxide in a gasogene about 30 per cent of the available calories.

At first sight this method of working seems irrational, but for obtaining high temperatures there are practical advantages, whose importance far exceeds the loss of heat in the gasogene. It permits much more elevated temperatures, and the recovery of a large portion of the heat, which in direct systems of heating in high temperature furnaces passes to the chimney as complete loss. There is actually an economy in the ordinary metallurgical methods even with this loss.

By means of gasogenes, we produce three kinds of gaseous fuel: the gas called *producer* or *air gas*, formed by the incomplete combustion of the fuel, with production of a mixed gas containing carbonic oxide and hydrogen compounds; the gas called *water gas*, from the decomposition of water by carbon at a high temperature, with production of carbonic oxide, hydrogen, and hydrogen compounds; and the gas called *mixed gas*, from the mixture of the two preceding ones by a process which combines the production of the two gases in the same furnace.

PRODUCER OR AIR GAS.

We have said that *air gas* results from incomplete combustion, and that its formation causes a loss of one third of the calories resulting from the complete combustion of the fuel. These gases contain, naturally, the nitrogen of the air used, to which must be added that of the air necessary to change the carbonic oxide and the hydrogen to carbonic acid and water.

The heat of combustion and the composition determined by different experimenters varies considerably, showing that they did not always work with average samples.

The proportion of nitrogen in these gases reaches 56 to 60 per cent; that of carbonic oxide, 21 to 32 per cent; that of of hydrogen, from traces to 17 per cent. The theoretical calculation for the combustion of carbon in air to a gas containing only carbonic oxide and nitrogen gives for the first 34.7 and for the second 65.3 per cent.

By adopting for the composition of air the round numbers 79 and 21, and for the weight of oxygen 1.430 grams per litre, for carbon the atomic weight of 12, and for oxygen 16,

$$12 : 16 = 1000 \text{ grams} : 1333 \text{ grams}.$$

A kilogram of carbon needs, then, 1⅓ kilograms of oxygen. A litre of oxygen weighing 1.430 grams, 1333 grams would occupy 932 litres. These 932 litres will give with carbon a double volume, or 1864 litres carbonic oxide. Multiplying 932 litres by the coefficient 4.77 (see Table XIV), we obtain the volume of the air corresponding, or 4445 litres. The gases of combustion will be composed then of these 4445 litres of air and the 932 litres of increase in volume, or 5377 litres for 1 kilogram of carbon. The 4445 litres of air will contain (at 79 per cent) 3513 litres of nitrogen, or 65.3 per cent.*

The calculation is more complicated when we have fuel containing hydrogen, as one portion of the oxygen disappears by its combination with the hydrogen to form water. Take for example, a coal containing 90 per cent of carbon, 5 per cent of hydrogen, and 5 per cent of oxygen. Suppose 1 kilogram of this coal, under theoretical conditions, burnt in a gasogene, i.e., with perfect transformation of the carbon into carbonic oxide and no residues. This coal contains 900 grams carbon, 50 grams hydrogen, 50 grams oxygen. 900

---

* One pound of carbon requires 1.333 lbs. of oxygen; 1 cubic foot of oxygen weighs 0.08926 lb.; 1.333 lbs. measure 14.93 cu. ft. These would give 29.86 of CO. 14.93 × 4.77 = 71.216, and 71.216 + 14.93 = 86.146, volume of gases of combustion. These contain 56.26 cu. ft. of nitrogen.

grams carbon produce 2100 grams carbonic oxide, requiring 1200 grams oxygen. 1200 grams oxygen occupy 839 litres. 50 grams hydrogen produce 450 grams water, and require 400 grams oxygen. These 400 grams oxygen occupy 279 litres. But the coal itself contains 50 grams oxygen, occupying 35 litres.

We have, then, $839 + 279 - 35 = 1083$ litres of oxygen required, and to calculate the amount of air needed multiply by 4.77. This gives 5163 litres of air needed for the incomplete combustion of 1 kilogram of carbon. These 5163 litres contain 4080 litres of nitrogen.

To obtain the total volume of gases produced by the incomplete combustion, we may add to the volume of the air introduced the volume due to the formation of carbonic oxide, and this is equal to the volume of the oxygen used, or 839 litres. We have, then, $5163 + 839 = 6002$ litres. But a quantity of oxygen has disappeared corresponding to the formation of the water, or $279 - 35 = 244$ litres (35 litres exists in the coal as above), and $6002 - 244 = 5758$ litres of gas produced by the incomplete combustion of 1 kilogram of coal.

Now, then, 5163 litres of air contain 4079 litres of nitrogen, which would form $\frac{4079}{5758}$, or 70.8 per cent of the total gas. All these numbers are at 0° and 760 mm. pressure.*

---

*One pound of coal would be 6300 grains carbon, 350 grains oxygen, and 350 grains hydrogen; 0.90 lb. carbon produces 2.1 lbs. carbonic oxide, and needs 1.2 lbs. oxygen; 1.2 lbs. oxygen occupies 13.44 cu. ft.; 0.050 lb. hydrogen produces 0.450 lb. water, and needs 0.400 lb. oxygen, or 4.48 cu. ft. The 0.05 lb. of oxygen in the coal occupies 0.56 cu. ft. Then $13.44 + 4.48 - 0.56 = 17.36$ of oxygen required $17.36 \times 4.77 = 82.81$ cu. ft. of air, containing 65.41 cu. ft. nitrogen. Total gases, $82.81 + 13.44 - 3.92 = 92.33$ total volume of gas, and

$$\frac{65.41}{92.33} = 70.8 \text{ per cent.}$$

Generally gasogenes contain less nitrogen, different causes producing diminution, among which are the use of a lower hydrogen coal than we have taken, and the decomposition of the fuel in the body of the furnace with a certain quantity of aqueous vapor formed during the combustion, or from the moisture in the air supplied.

Mahler determined the heat of combustion of a sample of gas from the Follembray glass-house, and found its composition per volume, using coal from Béthune, to be:

Marsh gas	2
Hydrogen	12
Carbonic oxide	21
Carbonic acid	5
Nitrogen	60
	100

The heat of combustion calculated from its composition is:

$$\text{Marsh gas} \quad 0.02 \times 10038 = 200.8$$
$$\text{Hydrogen} \quad 0.12 \times 3091 = 370.9$$
$$\text{CO} \quad 0.21 \times 3043 = 639.0$$
$$\overline{1210.7}$$

With the bomb he found 1212 calories.

WATER GAS AND MIXED GAS.

Water gas is produced when water is decomposed at high temperatures by fuels containing but little hydrogen, such as anthracite, charcoal, or coke. Mixed with hydrocarbon vapors, added to enrich it, or which may have been decomposed with the aqueous vapor, it serves for the illumination of a great number of cities, principally in America. But this is not its only use, as it is used for heating, and also for gas-engines. Mixed with producer gas, it has become a powerful means of heating, especially where high temperatures are wanted.

Water gas contains but little nitrogen: this is its main distinction from producer gas, and that which gives it a special value from an economical heating point of view.

We have previously stated (page 97) that during the combustion of carbon in a gasogene, there occurs a generation of nearly one third of the total heat were the fuel completely burnt. Besides this, the combustion produces a gas containing about one third its weight of combustible gas and two thirds inert gas (nitrogen), which is mixed with it.

These are important causes of two sources of loss in calories. In an air-gasogene one third of the calories is lost, since the gaseous products give up most of their sensible heat before being used. The 66 per cent of inert gas carries off an enormous quantity of heat to the chimney, and thence to the open air. It was with the idea of regaining or stopping these losses, or at least a large portion of them, that water gas originated.

Aqueous vapor and carbon, when submitted to a high temperature, produce carbonic oxide and hydrogen. Theoretically these are free from nitrogen; but there is always present a small percentage for various causes. In the air gasogene 12 kilogram of carbon and 16 kilograms of oxygen (atomic weights) unite to form 28 kilograms of carbonic oxide. On the other hand, 12 kilograms of carbon and 18 kilograms of water form 28 kilograms of carbonic oxide and 2 kilograms of hydrogen. Then 1 kilogram of carbon furnishes 2.5 kilograms of gas composed of carbonic oxide and hydrogen.

One kilogram of hydrogen has a caloric energy of 29042 calories.* These calories represent also the quantity of heat necessary to decompose the water; in the case of the water gas gasogene they are formed by the carbon burnt. The 12 kilograms of carbon will have to furnish, then, the calories necessary to decompose 18 kilograms of water; that is,

$$2 \times 29042 = 58084 \text{ calories.}$$

---

* Water being considered as vapor.

But 12 kilograms of carbon, in burning, generate only

$$12 \times 2473 = 29676 \text{ calories.}$$

To decompose the water, then, there is a shortage of force of

$$58084 - 29676 = 28408 \text{ calories}$$

for 2 kilograms of hydrogen, or 14204 calories for 1 kilogram. The heat must be furnished by an external source. In other terms, to gasify 1 kilogram of carbon there must be supplied

$$14204 \div 6 = 2367 \text{ calories.}$$

As may be easily seen, this operation absorbs much heat, and the combustion of the water gas can give only the calories used at first in forming it. The heat necessary for the decomposition of the water is actually taken from that of the preparatory period of the air gasogene, which makes a loss of one third of the total calories. In burning the water gas made under these conditions we utilize a part of the heat which would have been lost by the air gasogene only.

The decomposition of water by carbon is not as simple as would appear from the equation

$$H_2O + C = CO + H_2.$$

The lower portion of the fuel of the gasogene undergoes ordinary combustion on account of air being present; while in the upper portion the reaction takes place between the gaseous products formed in the lower portion and the heated carbon. The carbonic acid is then in contact with the heated carbon and is reduced to carbonic oxide:

$$C + CO_2 = 2CO.$$

Thus, the reaction with the water would be

$$5H_2O + 3C = 2CO_2 + CO + 10H;$$

carbonic acid being reduced to carbonic oxide in the final reaction, as in the case with the air gasogene.

Nine kilograms of aqueous vapor and 6 kilograms of carbon produce 1 kilogram of hydrogen and 14 kilograms of carbonic oxide, that is, a mixed gas is produced containing about one half its volume of each gas.

One cubic metre of hydrogen weighs 85.5 grams; one of carbonic oxide, 1194 grams. Then the volumes occupied by each gas would be 11.69 for hydrogen and 11.13 for carbonic oxide, or 51.23 per cent of hydrogen and 48.77 per cent of carbonic oxide.

From the foregoing account, it will be seen that the intermittent flow is a cause of great loss of caloric in the working of the water gasogene; but when a gas is wanted solely for heating at high temperatures, it may be obtained by a mixed system working continuously. The gasogene is filled with a mixture of air and steam, the air being employed in the proper proportion to keep up the heat necessary, or, in other words, to furnish by the combustion of part of the carbon, the number of calories necessary to the gasification of the other part.

We have seen (page 103) that to gasify 1 kilogram of carbon 2367 calories were needed. To maintain the heat this quantity must be produced by the action of the air. Mixed gases are poorer than water gas, as they contain more nitrogen and carbonic oxide and less hydrogen. Theoretically, we should attain the result of furnishing the heat to the gasogene necessary to maintain the temperature by supplying the steam sufficiently superheated; a gas very poor in nitrogen would then be made. But the superheating of steam causes new losses of heat.

NATURAL GAS.

Natural gas has been known for thousands of years in Asia, on the Caspian Sea, where it has long been a feature in religious services, but it is only recently that it has become of any use to man and played any part in the fuel world.

The natural gas output in the United States has attracted considerable attention since 1875, and especially since 1880. This gas always accompanies petroleum, although petroleum does not always accompany the gas. The wells are situated in various portions of New York, Pennsylvania, Ohio, Indiana, West Virginia, Kentucky, Tennessee, Colorado, California, and on the Canadian side also in numerous locations.

Natural gas is not of a constant or uniform composition, varying very much according to the locality from which it is taken. The individual constituent gases vary between wide limits, hydrogen at some places being almost wanting, while at others it is as high as 35 or 40 per cent. Marsh gas is in every case the principal constituent, but this runs down as low as 40 per cent in some analyses. Nitrogen is sometimes absent, and when present in large amounts, it is supposable that the gas analyzed was contaminated with atmospheric air.

The Ohio and Indiana fields yield gas of nearer a uniform composition than any of the others. The following table is typical:

	Ohio.			Indiana.		
	Fostoria	Findlay.	St. Mary's	Muncie.	Anderson	Kokomo.
Hydrogen	1.89	1.64	1.94	2.35	1.86	1.42
Marsh gas	92.84	93.35	93.85	92.67	93.07	94.16
Olefiant gas	0.20	0.35	0.20	0.25	0.47	0.30
Oxygen	0.35	0.39	0.35	0.35	0.42	0.30
Carbonic oxide	0.55	0.41	0.44	0.45	0.73	0.55
Carbonic acid	0.20	0.25	0.23	0.25	0.26	0.29
Nitrogen	3.82	3.41	2.98	3.53	3.02	2.80
Hydrogen sulphide	0.15	0.20	0.21	0.15	0.15	0.18

In addition to difference in composition in different localities, the composition of the gas varies considerably from time to time in each well. This is shown by the following analyses made at different times within a period of three months from a well at Pittsburgh, Pa.:

	1	2	3	4	5	6
Hydrogen	9.64	14.45	20.02	26.16	29.03	35.92
Marsh gas	57.85	75.16	72.18	65.25	60.70	49.58
Olefiant gas	0.80	0.60	0.70	0.80	0.98	0.60
Methane	5.20	4.80	3.60	5.50	7.92	12.30
Oxygen	2.10	1.20	1.10	0.80	0.78	0.80
Carbonic oxide	1.00	0.30	1.00	0.80	0.58	0.40
Carbonic acid	0.00	0.30	0.80	0 60	0.00	0.40
Nitrogen	23.41	2.89	0.00	0.00	0.00	0.00

The quantity of gas used daily in the town of Findlay, Ohio, in 1890, was estimated by Professor Orton to be, for

Glass-furnaces	10000000 cubic feet.
Iron mills	10000000 " "
Other factories	6000000 " "
Domestic use	4000000 " "
Total per day	30000000 " "

In Indiana, large wells have been opened and used as in Ohio. In Pennsylvania, several of the large rolling-mills and glass-houses near Pittsburg were formerly supplied with millions of feet per day; but the supply, used so lavishly, became exhausted. In Canada, at Fort Erie and Windsor are wells, the gas from which is piped across the river to Buffalo and Detroit respectively. All through the oil regions gas wells are to be found more or less, accompanying every well sunk.

From the composition of the gas, it will readily be seen that it is a valuable source of heat, the calorific power reaching 10000 calories or 1100 B. T. U. per cubic foot. It is used for domestic purposes, steam, glass making, iron mills, brick burning, and numerous other ways, and until recently used wastefully in all.

As compared with coal, 57.25 pounds of coal or 63 pounds of coke are about equal to 1000 cubic feet of the gas. The actual equivalent in steaming or furnace work varies with the furnace, and probably with the people using it. Equivalent values of 14000 to 25000 cubic feet per ton of coal are reported, and hardly any two users will give the same yield. It seems to be especially adapted to glass-making, giving a long, clean, ashless, smokeless flame, and hundreds of glass-pots were set up in the neighborhood of the wells, especially in Ohio. Each pot consumes from 58000 to 61000 cubic feet per 24 hours in window-glass works and from 31000 to 49000 cubic feet in flint-glass works, the difference being of course due to difference in burners and men, the gas being the same.

In all cases where this gas is used the chief claim made, in addition to those of gases generally, has been cheapness, and it has been sold without any regard to its actual value. A comparison of its value with that of other gases is given by McMillin in the Report of the Ohio Geological Survey, vol. VI, page 544, as follows:

1000 feet natural gas will evaporate.... 893 pounds of water.
  "   "   coal   "   "   " 591   "   "
  "   "   water   "   "   " 262   "   "
  "   "   producer gas "   " 115   "   "

OIL GAS.

There are several processes for producing gas from oil, usually petroleum or its derivatives. Some of them decompose the oil by means of heat alone, while others use steam, or steam and air together. The most successful pure oil process is the Pintsch; this is used extensively in the large cities of Europe and America to obtain a gas for illuminating cars on railways. The gas is made by allowing the oil to fall drop by drop on a strongly heated surface. Complete decomposition occurs, and a gas of high candle-power is formed. This is collected, and after compression supplied to the consumers. It loses some 20 per cent of the illuminating power during compression. As a source of heat, its use is, so far, very limited. An analysis and heat test will be found in the tables.

The Archer gas process is somewhat similar to the Pintsch, but the products of decomposition are generated at a comparatively low temperature, and then superheated subsequently so as to make the gas permanent. This gas is used for metallurgical purposes, but its use for heating boilers is very limited.

The other gases made with steam or steam and air have been advertised or pushed as fuel gases for several years. Many plants have been established and failed. A few of the most prominent are mentioned in the tables.

OTHER GASES.

Gas has been obtained from destructive distillation of wood, rosin, fats, and other materials. They were used principally for illumination, and seldom if ever for heat. They are now made only in very exceptional cases.

CHAPTER X.

CALORIFIC POWER OF COAL BURNT UNDER A STEAM-BOILER.

FUEL USED AND WATER EVAPORATED.

DISTRIBUTION OF THE HEAT PRODUCED.

EXPERIMENTS in heating steam-boilers have to determine:

1. How much water is vaporized by a given quantity of coal, so as to compare it with other coals or fuels;
2. The evaporative power of the steam-boiler used;
3. A comparison of the various styles of grates or methods of heating applied to steam-boilers.

In this book we will consider only the first case, the others being outside of its scope.

The knowledge of the heat of combustion of coal and other fuels is closely connected with experiments in heating steam-boilers. It is not enough to know the proportion of water which the apparatus or the fuel tested will vaporize: we must also determine the number of calories lost. We must know, besides, the composition of the coal and its heat of combustion, to determine the proportion of calories used to that possible with perfect combustion.

The first work in this direction worth mentioning was probably that done by Peclet in 1833, but his results were very crude, and are of no account now. The next were those made by Prof. Johnson, in 1842 and 1843, for the U. S. Navy Department, to determine the steaming powers of the coals then in use. He analyzed and tested some thirty-five different coals, domestic and foreign. The tests were made with a specially built boiler, and careful and copious notes were taken all through. The chimney gases were analyzed, and an attempt made to determine their quantity. In 1891 Mr. W. Kent[*] reviewed his work, and found that, with corrections for the constants employed by Johnson, the tests were comparable with those made at the present time. The figures given in the tables as Johnson's are with Kent's corrections.

The first experiments based on the knowledge of the composition and heat of combustion of coal were published in 1868 and 1869 in the *Bulletin de la Société Industrielle de Mulhouse*. Scheurer-Kestner remarks in the first part of this work, which he prosecuted later on with assistance of Meunier-Dollfus (*loc. cit.* p. 1):

"It is necessary to analyze the great difference found between the theoretical heat of combustion (at that time no actual determinations had been made) and the practical yield.

"Several elements of the calculation aid in making this shortage. The principal ones are:

"The heat of combustion of the coal;

"The composition of the coal;

"The composition of the cinders as drawn from the ash-pit;

"The quantity of water vaporized and the temperature of the steam produced;

"The volume of gases introduced under the grate, and their temperature when they leave the boiler to pass into the chimney;

"The composition of the gaseous products of combustion;

---

[*] Engineering and Mining Journal, Oct. 1891.

"The temperature of the cinders at the time of dumping;

"The loss of caloric by radiation from the setting of the boiler."

We must refer to mineral and organic as well as gas analysis to obtain the necessary elements for the distribution of the caloric produced by the combustion of the coal on a steam-boiler grate.

To avoid referring to them, we will consider the composition and heat of combustion of coal as known. (See tables.)

WEIGHT OF FUEL.

The coal used in the test should be kept under cover away from moisture and heat, so that the hygroscopic water it contains shall vary as little as possible from the time of taking the sample. Weigh the coal in the gross, and then weigh portions of about 100 kilograms (220 lbs.) on a scale sensible to $\frac{1}{1000}$.

Where practicable, a box open at the top and holding 500 pounds of coal should be provided for each 25 square feet grate area, and in proportion for larger grates. It should be placed on the scales, and conveniently located for shoveling into the fire.

The exact time of weighing should be noted and the exact weight set down. The weight should be taken at the instant of closing the fire-door. The box should be completely emptied each time. The difference of weight at each firing will give the several quantities fired; the differences of time will give the intervals between firing; and the difference of time between successive charges will serve as a check on the record of the test. A chart or diagram should be made showing the regularity of the working, and it is well to keep the records in tabular form; weights in one column, time in another.

SAMPLING THE COAL.

In all experiments for determining heat of combustion of fuels, the sampling must be done with the utmost care, especially if the laboratory and working test are to be made at the same time. Samples accurately representing the coal of the working test must be kept in the laboratory, and when coal is tested which contains foreign matter and considerable moisture, too much care cannot be taken to prevent errors.

The official method of the American Society of Mechanical Engineers is given in the Appendix, and answers the purpose very well. If very large quantities are to be sampled, remove a portion from each cart-load and then re-sample these as per directions above mentioned.

It is not always necessary to resort to these methods. When the coal comes from the same pit and level, experience has shown that a piece which seems to agree with the general character is usually sufficient. Care must be taken to avoid samples having too much hanging-wall or bed-rock. For twenty years the pure coal of Ronchamp taken from the same pit has given the same calorimetric test, when it contained from 10 to 20 per cent of ash. Lord and Haas[*] showed that the same was true of many American mines, especially in Ohio and Pennsylvania. This being true, we could consider that in sampling we did not sample the coal, but the impurities; and that a sample showing the average impurities would give all that was needed, as we would know what the coal was.

Care must be taken with regard to the moisture, and any coal showing much external moisture must be examined as near as possible to the original condition. For example, a coal containing 10 per cent of moisture in the pile may, after sampling, crushing, and resampling, lose all but 4 or 5 per cent. If the moisture was determined in this coal while in as

---

[*] Trans. Am. Inst. Min. Eng., Feb. 1897.

large pieces as possible, this moisture would all be accounted for.

In spite of all precautions, samples do not always agree in mineral content with the mass. The difference seems to be due not only to the unequal distribution of the foreign mineral matter throughout the coal, but principally to the difference in specific gravity between the coal and this mineral, so that the purer the coal the more satisfactory the sampling.

Sometimes a coal is rich in foreign matter, and is contained in a tube open at one end. From this samples may be drawn showing differences of several per cents; as for example, 12.49 and 16.74 per cent obtained in two successive cases. The following experiment shows how this happens and how to prevent it: 30 grams of coal, finely pulverized, and containing 20 per cent of mineral, was put into a glass tube, which was closed with a cork and placed vertically, giving it slight taps to settle it down. In a short time most of the foreign material was at the bottom of the tube, the upper portion being nearly free. To avoid such an error the sample must be drawn only after thorough mixing, and without any shaking or jarring of the tube. It is well to use pastilles made up immediately after thorough mixing. A sample containing only 13 to 14 per cent of foreign matter has given from a tube, 12.20, 12.81, 13.12, 13.50, 14.42 per cent.

ANALYSIS OF THE COAL.

No attempt will be made to treat the methods of analyzing coal; still, as this usually accompanies a calorimetric determination, some hints may be useful. Scheurer-Kestner usually burns the coal in tubes of white glass placed on an iron gutter. The same tube may thus serve several times if asbestos cloth be placed between the tube and the iron and the cooling be properly regulated. His tubes are 70 to 75 centimetres (27 to 29 inches) long and 15 to 20 millimetres

(0.6 to 0.8 inch) inside diameter. They are filled with copper oxide in small pieces, except at the front end, which has a small piece of metallic copper, and at the back, where the platinum boat containing the coal is placed. Usually half a gram is used for a test, the coal having been previously dried at 100° to 105° C. (212° to 221° F.).

Before putting in the sample the tube is heated to redness and thoroughly dried by means of a current of dry oxygen. The combustion is carried on so as to allow time enough for all the gas to be absorbed by the potash, during the first half of the time the bubbles passing through very slowly. There is no risk then of unburnt gases passing off. An iron or a platinum tube may be used in place of the glass one, but glass allows inspection at all times.

An analysis should show the carbon, hydrogen, oxygen, nitrogen, sulphur, ash, and moisture, and they should be so given that the carbon, hydrogen, oxygen, nitrogen, sulphur, and ash should equal 100 per cent, the moisture being determined separately, or if preferred all but ash and moisture may foot up 100, and those two be given separately. This latter method is the one which is followed by many of the European engineers, and will be found so in the tables given at the end of this book. If possible the approximate analysis should also be given.

In determining the moisture too much care cannot be taken to expel all of it. With many coals, and especially our Western ones, the ordinary heating to 110° C. is not sufficient. Kent, Carpenter, Hale, and others have investigated this question, and find that a much higher temperature is needed, and must be employed. In some cases as high as 140° to 150° C. may be used with safety, and such temperatures are recommended by Carpenter, no appreciable amount of volatile matter being driven off.

ANALYSIS OF THE CINDERS.

The cinders and ashes produced by the combustion of the coal are collected so as to weigh and sample them. After drying and determining the water the sample is put into a glass tube as with coal. As the quantity of hydrogen is usually very small, it need not be determined, and the calcination for the carbon can be performed in the open air. The following table contains the results of the tests made by Scheurer-Kestner and Meunier-Dollfus on steam-boiler cinders:

	1	2	3	4
Carbon	9.20	12.65	6.73	8.92
Hydrogen	0.37	0.29	0.21	0.27
Ash	89.95	86.50	92.64	91.42
	99.52	99.44	99.58	99.61

The proportion of carbon in cinders may be as low as 7 per cent, but is usually higher, and 10 to 12 per cent may be called good practice.

DURATION OF THE TEST.

A test should continue at least a whole day on account of certain irregularities and causes of error which are constant. The level of the water should be the same at the end of the test as at the beginning, since a slight difference in level means considerable water.

The condition of the combustion at the time of stopping cannot always be ascertained, and this produces a cause of uncertainty. Another cause is from the temperature of the water in the boiler, and especially in the economizer. On short runs these sources of error cause very faulty results.

THE WATER EVAPORATED.

The feed-water is preferably held in a gauged reservoir, or else weighed, meters not being certain unless checked frequently. Use only cold water or water whose temperature will vary but little during the test, so as to avoid corrections of temperature and expansion. The temperature usually varies so little that no account of this variation need be taken. Pump to the boiler with as much regularity as possible, and keep accurate record.

To have the same level at the end as at the beginning, keep up the initial pressure and feed very carefully. The mean temperature of the feed-water is referred to 0° C., considering that the specific heat is constant. Otherwise we may use Regnault's formula,

$$Q = t - 0.00002 t^2 + 0.0000003 t^3.$$

But when the temperature of the water varies no more than 10 degrees, no appreciable error will be made by calling t equal to the temperature.

TEMPERATURE OF THE STEAM.

We may measure the temperature of the steam directly by a thermometer in the boiler, or indirectly by observing the pressure. Both methods should be used.

To take the temperature directly, the thermometer is placed in an iron tube closed at one end and reaching to the middle of the boiler. The tube should be filled with paraffin or some analogous substance. The temperature of the steam or the water may be taken as desired by changing the position of the thermometer in the tube. See Figure 39. Vertical maximum and minimum thermometers are very useful, preventing too hasty observations.

To measure the temperature by pressure an air-thermometer is used. A registering manometer aids the work considerably, as observations should be taken regularly at frequent and equal intervals. The temperature is calculated by means of tables of vapor-tension.*

MOISTURE IN THE STEAM.

The percentage of moisture should be ascertained by means of a throttling or a separating calorimeter, directions for the use of which will be furnished by the makers. They should easily and completely separate the water in a manner convenient for measuring, or better, for weighing. It is advisable to use two or three at the same time, thus serving as checks for each other.

"The throttling steam-calorimeter was first described by Professor Peabody in the *Transactions*,† vol. X. page 327, and its modifications by Mr. Barrus, vol. XI. page 790; vol. XVII. page 617; and by Professor Carpenter, vol. XII. page 840; also the separating-calorimeter designed by Professor Carpenter, vol. XVII. page 608. These instruments are used to determine the moisture existing in a small sample of steam taken from the steam-pipe, and give results, when properly handled, which may be accepted as accurate within 0.5 per cent (this percentage being computed on the total quantity of the steam) for the sample taken. The possible error of 0.5 per cent is the aggregate of the probable error of careful observation, and of the errors due to inaccuracy of the pressure-gauges and thermometers; to radiation; and, in the case of the throttling-calorimeter, to the possible inaccuracy of the figure 0.48 for the specific heat of superheated steam, which

---

* For full details regarding setting up an open-air manometer, see paper by Scheurer-Kestner and Meunier-Dollfus in the *Bulletin de la Société industrielle de Mulhouse*, 1869, page 241; also *Trans. A. S. M. E.*, vol. VI. pages 281 and 282.

† Transactions A. S. M. E.

is used in computing the results. It is, however, by no means certain that the sample represents the average quality of the steam in the pipe from which the sample is taken. The practical impossibility of obtaining an accurate sample, especially when the percentage of moisture exceeds two or three per cent, is shown in the two papers by Professor Jacobus in *Transactions*,* vol. XVI. pages 448, 1017.

"In trials of the ordinary forms of horizontal shell and of water-tube boilers, in which there is a large disengaging surface, when the water-level is carried at least 10 inches below the level of the steam outlet, and when the water is not of a character to cause foaming, and when in the case of water-tube boilers the steam outlet is placed in the rear of the middle of the length of the water-drum, the maximum quantity of moisture in the steam rarely, if ever, exceeds two per cent; and in such cases a sample taken with the precautions specified in article XIII. of the Code may be considered to be an accurate average sample of the steam furnished by the boiler, and its percentage of moisture as determined by the throttling or separating calorimeter may be considered as accurate within one half of one per cent. For scientific research, and in all cases in which there is reason to suspect that the moisture may exceed two per cent, a steam-separator should be placed in the steam-pipe, as near to the steam outlet of the boiler as convenient, well covered with felting, all the steam made by the boiler passing through it, and all the moisture caught by it carefully weighed after being cooled. A convenient method of obtaining the weight of the drip from the separator is to discharge it through a trap into a barrel of cold water standing on a platform scale. A throttling or a separating calorimeter should be placed in the steam-pipe, just beyond the steam-separator, for the purpose of determining, by the sampling method, the small percentage of moisture which may still be in the steam after passing through the separator.

* Transactions A. S. M. E.

"The formula for calculating the percentage of moisture when the throttling-calorimeter is used is the following:

$$w = 100 \times \frac{H - h - k(T - t)}{L},$$

in which w = percentage of moisture in the steam, H = total heat and L = latent heat per pound of steam at the pressure in the steam-pipe, h = total heat per pound of steam at the pressure in the discharge side of the calorimeter, k = specific heat of superheated steam, T = temperature of the throttled and superheated steam in the calorimeter, and t = temperature due to the pressure in the discharge side of the calorimeter, = 212° Fahr. at atmospheric pressure. Taking $k = 0.48$ and $t = 212$, the formula reduces to

$$w = 100 \times \frac{H - 1146.6 - 0.48(T - 212)}{L}.\text{"}*$$

CORRECTIONS FOR QUALITY OF STEAM.[†]

Given the percentage of moisture or number of degrees of superheating, it is desirable to develop formulæ showing what we have termed "the factor of correction for quality of steam," or the factor by which the "apparent evaporation," determined by a boiler-test, is to be multiplied to obtain the "evaporation corrected for quality of steam." It has been customary to call the proportional weight of steam in a mixture of steam and water "the quality of the steam," and it is not desirable to change this designation. The same term applies when the steam is superheated by employing the "equivalent evaporation," or that obtained by adding to the actual evaporation the

---

* William Kent in the Report of the Committee on Boiler-tests, A. S. M. E., 1897.

† C. E. Emery in the Report of Committee on Boiler-tests, A. S. M. E., 1897.

proportional weight of water which the thermal value of the superheating would evaporate into dry steam from and at the temperature due to the pressure. "The factor of correction for quality of steam" in a boiler-test differs from the "quality" itself, from the fact that the temperature of the feed-water is lower than that of the steam.

Let

$Q =$ quality of moist steam as described above;
$Q_1 =$ the quality of superheated steam as described above;
$P =$ the proportion of moisture in the steam;
$k =$ the number of degrees of superheating;
$F =$ the factor of correction for the quality of the steam when the steam is moist;
$F_1 =$ the factor of correction for the quality of the steam when the steam is superheated;
$H =$ the total heat of the steam due to the steam-pressure;
$L =$ the latent heat of the steam due to the steam-pressure;
$T =$ the temperature of the steam due to the steam-pressure;
$T_1 =$ the total heat in the water at the temperature due to the steam-pressure;*
$J =$ the temperature of the feed-water;
$J_1 =$ the total heat in the feed-water due to the temperature.*

Therefore, for moist steam,

$$Q = 1 - P, \quad \cdots \cdots (1)$$
$$P = 1 - Q, \quad \cdots \cdots (2)$$
$$Q + P = 1. \quad \cdots \cdots (3)$$

See also equation (6).

---

* Most tables of the properties of steam and of water are based on the total heat of steam and water above 32 degrees Fahr. For such tables the total heat in the water at a given temperature is equal approximately to the corresponding temperature minus 32 degrees. Exact values should, however, be taken from the tables.

With both the condensing and throttling calorimeters the water and steam are withdrawn from the boiler at the temperature of the steam, and with a separator the water can only be accurately measured when under pressure, so that the difference between the steam and the moisture in the steam, as they leave the boiler, is simply that the former has received the latent heat due to the pressure, and the latter has not. There is, however, imparted to the water in the boiler not only the latent heat in the portion evaporated, but the sensible heat due to raising the temperature of all the water from that of the feed-water to that of the steam due to the pressure.

In equation (3) the proportional part Q receives from the boiler both the sensible and the latent heat, or the total heat above the temperature of the feed $= Q(H - J_1)$ thermal units, and the part P the difference in sensible heat between the temperatures of the steam and of the feed-water $= P(T_1 - J_1)$ thermal units. If all the water were evaporated, each pound would receive the total heat in the steam above the temperature of the feed, or $H - J_1$. "The factor of correction for the quality of the steam," when there is no superheating, is therefore

$$F = \frac{Q(H - J_1) + P(T_1 - J_1)}{H - J_1} = Q + P\left(\frac{T_1 - J_1}{H - J_1}\right). \quad . \quad (4)$$

The superheating of the steam requires 0.48 of a thermal unit for each degree the temperature of the steam is raised, so for k degrees of superheating there will be $0.48k$ thermal units per pound weight of steam, and the "factor of correction for the quality of the steam" with superheating.

$$F_1 = \frac{H - J_1 + 0.48k}{H - J_1} = 1 + \frac{0.48k}{H - J}. \quad . \quad . \quad (5)$$

See also equation (7).

With the throttling-calorimeter the percentage of moisture P, or number of degrees of superheating, are determined as explained before.

Since the invention of the throttling-calorimeter the use of the original condensing, or so-called barrel, calorimeter is no longer warranted. Accurate results should, however, be obtained by condensing all the steam generated in the boiler, and this plan has been followed in certain cases. It has, therefore, been thought desirable to add other formulæ applicable to condensing-calorimeters. The following additional notation is required:

$W =$ the original weight of the water in calorimeter, or weight of circulating water for a surface condenser.

$w =$ the weight of water added to the calorimeter by blowing steam into the water, or of "water of condensation" with a surface condenser.

$t =$ total heat of water corresponding to initial temperature of water in calorimeter.

$t_1 =$ total heat of water corresponding to final temperature in calorimeter.

Evidently, then:

$W(t_1 - t) =$ the total thermal units withdrawn from the boiler and imparted to the water in calorimeter.

$\dfrac{W}{w}(t_1 - t) =$ the thermal units per pound of water withdrawn from the boiler and imparted to the water in calorimeter, from which should be deducted $T_1 - t_1$ to obtain the number of thermal units per pound of water withdrawn from the boiler at the pressure due to the temperature T.

Since only the latent heat L is imparted to the portion of the water evaporated, the quality Q, or proportional quantity evaporated, may be obtained by dividing the total thermal units per pound of water abstracted at the pressure due to the temperature T by the latent heat L. Hence, as given in

Appendix XVII., 1885 Code, with some differences in notation,

$$Q \text{ and } Q_1 = \frac{1}{L}\left[\frac{W}{w}(t_1 - t) - (T_1 - t_1)\right]. \quad . \quad . \quad (6)$$

The value Q applies when the second term is less than unity. P may be derived therefrom by substitution in equation (2) and F from equation (4).

Q_1 applies when the second term of the above equation is greater than unity, which shows that the steam is superheated, and, as in this case, the heating value of the superheat has already been measured by heating the water of the calorimeter; the proportional thermal value of the same, in terms of the latent heat L, is represented directly by $Q_1 - 1$, and we have as the factor of correction for the quality of the steam with superheating,

$$F_1 = \frac{H - J_1 + L(Q_1 - 1)}{H - J_1} = 1 + \frac{L(Q_1 - 1)}{H - J_1}. \quad . \quad (7)$$

See also equation (5).

When the quality is greater than 1, or equals Q_1, the number of degrees of superheating,

$$k = \frac{L(Q_1 - 1)}{0.48} - 2.0833 L(Q_1 - 1). \quad . \quad . \quad (8)$$

THE QUALITY OF SUPERHEATED STEAM.*

The quality of the superheated steam is determined from the number of degrees of superheating by using the following formula:

$$Q = \frac{L + 0.48(T - t)}{L},$$

---

* G. H. Barrus in Report of Committee on Boiler-tests, A. S. M. E., 1897.

in which L is the latent heat in British thermal units in one pound of steam of the observed pressure; T the observed temperature, and t the normal temperature due to the pressure. This normal temperature should be determined by obtaining a reading of the thermometer when the fires are in a dead condition and the superheat has disappeared. This temperature being observed when the pressure as shown by the gauge is the average of the readings taken during the trial, observations being made by the same instrument, errors of gauge or thermometer are practically eliminated.

CHAPTER XI.

AIR SUPPLIED AND GASEOUS PRODUCTS OF COMBUSTION.

VOLUME OF AIR NECESSARY TO COMBUSTION.

Four elements are to be considered in calculating the theoretical volume of air for combustion: carbon, hydrogen, oxygen, sulphur. The last is sometimes wanting in coal, but not usually.

Carbon.—The atomic weights of carbon and oxygen are as 12 and 16, and 2 atoms of oxygen are needed to form carbonic acid with 1 atom of carbon. Then

$$12 : 32 = 1 : 2.666.$$

1 kilogram of oxygen occupies 0.699 cubic metre (Table IV); 1 kilogram of carbon needs

$$0.699 \times 2.666 = 1.863 \text{ cubic metres of oxygen.}$$

Hydrogen.—The atomic weights of hydrogen and oxygen being respectively 1 and 16, and water being formed of 2 atoms of hydrogen and 1 of oxygen, we have

$$2 : 16 = 1 : 8;$$

and as 1 kilogram of oxygen occupies 0.699 cubic metre, 1 kilogram of hydrogen requires

$$8 \times 0.699 = 5.592 \text{ cubic metres of oxygen.}$$

Sulphur.—The atomic weights of sulphur and oxygen being as 32 to 16, and sulphurous acid containing 1 atom of sulphur and 2 atoms of oxygen, we have

$$32 : 32 = 1 : 1.$$

1 kilogram of oxygen occupies 0.699 cubic metre; 1 kilogram of sulphur needs, then, to form sulphurous acid

$$1 \times 0.699 = 0.699 \text{ cubic metre of oxygen.}$$

As most fuels have some oxygen in their composition, we must deduct this at the rate of 0.699 cubic metre per kilogram.

Then multiplying these results by 4.77 (Table XIV) we obtain the number of cubic metres of air required.

A similar method of calculation will give

For one pound of carbon....... 29.86 cubic feet of oxygen.
" " " " hydrogen .... 89.60 " " " "
" " " " sulphur...... 11.20 " " " "

As an example, take a coal containing 90% C, 5% H, 3.5% O, 0.1% N, and 0.5% S.

$$
\begin{aligned}
\text{C} &\ldots\ldots\ 0.900 \times 1.863 = 1.677 \text{ cubic metres.} \\
\text{H} &\ldots\ldots\ 0.040 \times 5.592 = 0.224 \\
\text{S} &\ldots\ldots\ 0.005 \times 0.699 = 0.003 \\
&\qquad\qquad\text{Total oxygen}\ldots\ldots\ldots 1.904 \\
\text{O} &\ldots 0.035 \times 0.699 = 0.024 \\
&\qquad\qquad\qquad\qquad\qquad\quad 1.880
\end{aligned}
$$

$1.880 \times 4.77 = 8.967$ cubic metres of air per kilogram of coal; or 143.98 cubic feet of air to the pound of coal.

This result of course is only approximate, as complete combustion is not attained with coal and solid fuels. With liquid fuels, and especially gases, however, the combustion is usually complete.

Tables V and VI gives the coefficients to be employed in the calculations.

Table XIII gives the theoretical quantity of air required for the combustion of various fuels; the actual quantity used depends on the conditions of firing, fuel, etc, and is seldom less than twice the amount shown in the table, except perhaps with gases.

VOLUME OF WASTE GASES BY ANALYSIS.

For a long time efforts have been made to determine the quantity of air used by comparison of the analyses of the waste gases with those of the fuel used. Many analyses have been published, but the results showed so little regularity, and were so contradictory even, that it was impossible to form any conclusion further than that waste gases from coal may contain at the same time both combustible gas and an excess of air.

Peclet, in 1827, published the first analyses, made with samples collected from a boiler-stack by means of an inverted flask containing water. Ebelmen, in 1844, published a memoir on the composition of gases from industrial furnaces. He analyzed the gases from a metallurgical furnace, the gas being collected by an aspirator. In 1847 Combes made a report on methods of burning or preventing smoke, giving analyses by Debette. In these the first attempts were made to obtain average samples, they being drawn at certain determined stages of the heat and the fuel.

In 1862 Commines de Marcilly published analyses of gases from locomotives, as well as from stationary boilers, but the author said the time of collection lasted only a few seconds. In 1866 Cailletet showed that, to obtain correct results, the gas should not be collected till somewhat cooled; otherwise, on account of dissociation, a larger proportion of combustible gas is found than when cooler.

But, on account of the defective methods of sampling

used, no conclusion other than that stated above can be drawn from these analyses, and no possible idea can be deduced as to the actual composition of the gases as a whole. When we try to use laboratory methods of control in practical workings, the first necessity is to obtain correct samples for analysis, that is, average samples. In this respect all the above-quoted authors are deficient. The tests made by Scheurer-Kestner, published in 1868, were the first to conform to this requirement. His samples were drawn by a system analogous in principle to that described for sampling coal.

It is not always necessary to resort to such a complicated operation in case of a permanent gas; samples taken from the general current by means of an ordinary aspirator or an oil-aspirator (page 132) will usually do if drawn at a sufficient distance from the fire. If the gases have passed through a long flue, especially one with several bends, they are sufficiently mixed, and may be considered as a homogeneous gas. We must remember, however, that as we recede from the fire the infiltration of air, if not prevented, becomes greater. In careful experiments, the method to be described of fractionating a large volume is preferable.

GAS SAMPLER.

In principle the apparatus consists of a falling-water aspirator, and a second mercury aspirator drawing a small fraction of the gases from the current of the first in a constant regular manner and keeping it in a mercury gas-holder, A (Fig. 28), which is a strong glass flask of 3 litres capacity, holding about 40 kilograms (88 lbs.) of mercury. The gas-holder is connected by the tube a with the tube c for sampling the gas, the flask A and its accessories acting as a Mariotte flask. It is closed at the top by a stopper hollowed out conically below and having holes for two tubes, a and b. This hollowing is to permit filling without any air-bubbles. The tubes *a* and *b* have glass stop-cocks, but the one in *a* may be omitted. The manometric tube *c* shows the pressure. Tube *d*, like *c*, passes through a rubber stopper, closing the horizontal tubulature of the gas-holder.

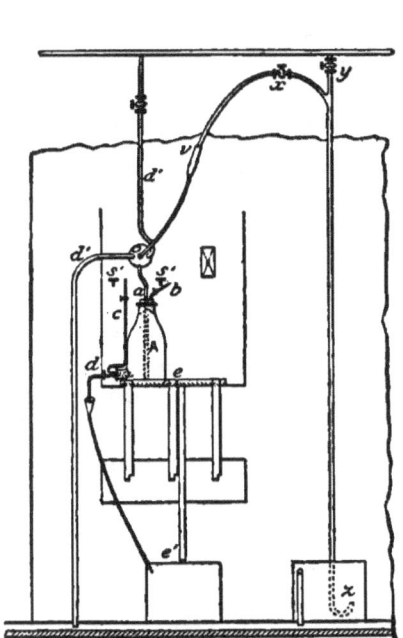

FIG. 28.—GAS SAMPLER.

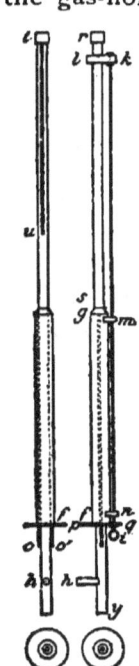

FIG. 29.—SAMPLER TUBE.

This tube can be rotated in the stopper to the position shown, or to one 180° from such position. The flask is graduated on the side into millimetres. Tube *a* fits the hole of the stopper tightly, and can be moved up or down as desired to suit the quantity of gas in the flask. All joints are covered with paraffin, tube *a* being greased to facilitate movement.

Fig. 29 shows the gas sampling tube. It consists of a platinum cylinder, *rs*, 10 millimetres (0.4 inch) diameter and 700 millimetres (27.5 inches) long, having a longitudinal slot of several centimetres length. The end *r* is closed with a platinum cap; the end s is soldered to a copper tube, sy, passing into a Liebig condenser having two tubes, oo', for the water. In most cases the platinum tube may be replaced without trouble by one of copper, or even iron, the platinum being necessary only when the gases are drawn at a temperaperature high enough to cause oxidation of the other metals. With iron or copper a portion of the oxygen is removed in the passage through the tube.

The tube ry is open at y, and has a side tube h. Aspiration is carried on through the opening in the platinum tube. A movable rod, ik, carrying a platinum scraper is attached to one end of the tube, and moves in the slot to clean it, as occasion requires, from soot, etc. The disk p serves to hold the cement used in fastening it to the stack or chimney, and prevents ingress of external air. The rod mn passes through a caoutchouc bearing fastened between the disks p and q.

Fig. 28 represents a front view of the apparatus. Fig. 30 represents a side view in elevation. The tube ry is introduced through an opening made for the purpose in the masonry, the part rs being exposed inside. The end y, is connected with a lead pipe, v, by a rubber tube; this pipe is soldered to another one, yz. On opening the cock y, water flows from a reservoir and empties at z. Suction in yrs should amount to several millimetres of mercury, and is regulated by the cocks y and x controlling the water-flow, and also by the length of yz. The gas drawn in by yvx may be measured by collecting it at z, and should amount to 4 or 5 litres (25 to 30 cubic inches) per minute.

The gas-holder is supported by a piece of sheet iron with upturned edges forming a shelf. Any mercury spattered over or spilled is thus easily collected. The mercury tank is supported from the wall of the chimney in such position as to facilitate refilling the flask through a siphon. The tubes dd' serve to feed the condenser.

While the current is passing through yr a small quantity

is drawn out by the tube h, and this should be so regulated by the cock d that only from $\frac{1}{350}$ to $\frac{1}{500}$ is collected.

Whenever the level of the mercury lowers, it shows a

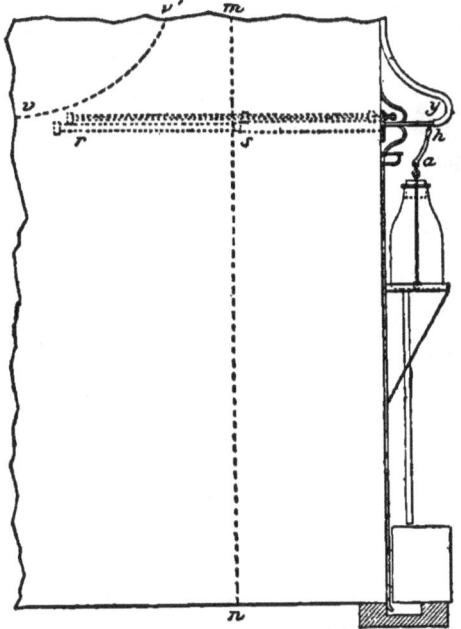

FIG. 30.—GAS SAMPLER.

clogging in the slot, and it should be cleaned by moving the rod. This always indicates when cleaning is necessary, and it sometimes keeps clean for hours.

When a sufficient sample has been obtained turn up the tube d, and then the gas-holder can be carried away.

The method recommended by the American Society of Mechanical Engineers is to have a "box or block of galvanized sheet iron equal in thickness to one course of brick," and secure in it a series of ¼-inch gas-pipes, all alike at the ends and of equal lengths, in such manner that the open ends may be evenly distributed over the area of the flue A (Fig. 32), and their other open ends enclosed in the receiver B.

If the flue-gases be drawn off from the receiver B by four tubes, CC, into a mixing-box, D, beneath, a good mixture can be obtained. Two such samplers, one above the other, a foot apart, in the same flue will furnish samples of gases which show the same composition by analysis.

The *oil gas holder* (Fig. 31) consists of a bottle tubulated at the bottom and connected with the supply of gas at the upper opening. It may contain some 10 litres (600 cubic inches), and is filled with water having on it a layer of 10 centimetres (4 inches) of oil. The water running out from the tubulature at the bottom draws the gas in at the top. The stopper at the top has two openings, through one of which passes a funnel-tube, through which water may be poured to expel the gas when portions of it are needed. The gas then passes out by the same tube through which it was drawn into the bottle.

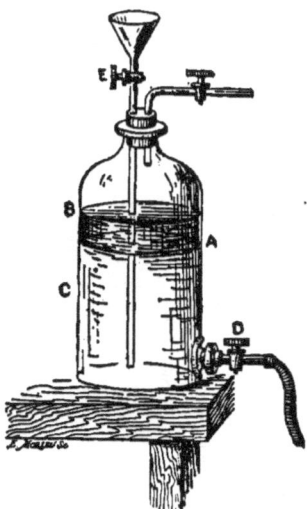

FIG. 31.—OIL ASPIRATOR.

With all kinds of aspirators or gas holders especial care must be taken to prevent entrance of air into the flue after leaving the fire, since the correct analysis will show not only the quantity of unburnt gases, but also the excess of air, and any mixture of outside air will vitiate the result and cause faulty deductions as to the working of the fire; and consequently the waste calories.

To prevent this, all joints in the masonry must be examined and repaired if necessary. In case of dampers, which must be used, the bearings can be made in stuffing-boxes, as recommended by Burnet. Generally, the gas can be sampled before it arrives at a damper, as the course of the boiler-flue is usually sufficient to cause a thorough mixing of the gases. In case there are several dampers, the first one may be dispensed with for the time being.

When the gases are taken quite near the fire, they must be drawn very slowly in order to gradually cool them down and

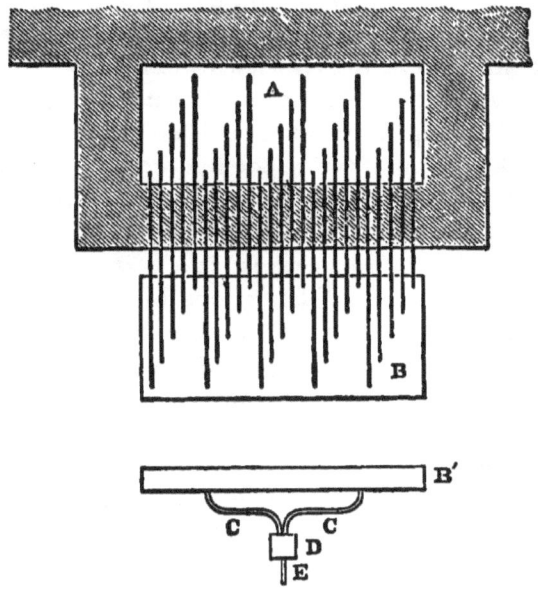

FIG. 32.

avoid dissociation. In this case a stoneware tube may be used for suction. If this precaution is neglected the gases collected may be entirely different from those passing off at the chimney. Metal tubes are inadmissible, since they abstract oxygen, and hence cause a change in composition.

ANALYSIS OF THE GASES.

The collected gases contain nitrogen, oxygen, carbonic acid, carbonic oxide, hydrocarbons, and occasionally free hydrogen. To determine all these a eudiometric method must be used; but usually only the oxygen, carbonic oxide, and carbonic acid are required. In normal combustion with sufficient air the quantity of hydrocarbons is very trifling, and need not be considered. This occurs usually with a supply of 15 cubic metres of air per kilogram (240 cubic feet per pound) of coal, and should produce a waste gas containing 10 to 14 per cent of carbonic acid, in which case the unburnt hydrocarbons amount to less than 1 per cent.

The Orsat apparatus or its modifications may be used to determine the oxygen, carbonic acid, and carbonic oxide. By using Winckler's modification the hydrocarbons may be determined. For exact analyses of the gases the Hempel apparatus may be used. For general work, however, the Orsat apparatus or the Orsat-Muencke is the best and most easily transported and handled. Directions for using this apparatus need not be given here, as they can be found in all works on gas analysis, or can be had of the dealers.

The following table gives analyses made by Scheurer-Kestner of waste gases from Ronchamp coal. The gases for examination were collected by means of the apparatus described above (pp. 128 *et seq.*) and shows the average for a whole day's run.

Air in Excess.	Percentage Composition of the Gases.						Coal per Hour per Square Foot of Grate.	Weight of Charge of Coal.	Frequency of Charging.
	Nitrogen.	Carbonic Acid.	Oxygen.	Carbonic Oxide.	Hydrocarbons.				
					Carbon.	Hydrogen.			
							Lbs.	Lbs.	
6.60	80.38	14.87	1.41	0.84	1.15	1.35	8.19	15.4	5'
10.47	80.60	14.16	2.18	0.97	0.98	1.11	9.625	30.8	8'
13.32	80.66	14.63	2.80	0.86	0.49	0.56	9.625	15.4	4'
17.61	81.52	13.34	3.77	0.86	0.46	0.91	8.19	15.4	3'
20.94	80.23	13.43	4.42	0.24	0.32	1.41	8.19	30.8	10'
26.18	80.34	12.89	5.53	0.24	0.28	0.96	4.71	15.4	8'
42.84	79.76	10.87	8.99	0.24	0.19	0.19	18.94	15.4	2'
53.78	79.86	8.23	11.35	0.24	0.04	0.52	3.41	13.2	10'

The following table gives some analyses by Bunte of gas samples from coal burnt in his experimental apparatus at Munich:

	Min. and Max. of Air.	CO_2	CO	H	O	N
Coal from the Ruhr............	....	10.26	0.53	0.01	10.00	79.20
Do.	....	16.45	1.94	1.45	1.52	78.64
Do.	....	13.40	0.48	0.30	6.52	79.30
Do.	....	11.45	1.22	0.78	7.27	79.28
Do. (grate more open).	....	8.15	0.10	0.01	11.60	80.14
Do. Do.	....	6.12	0.89	0.10	14.21	78.68
Coal from Saarbruck: Kœnig..	Min.	15.12	1.09	1.02	2.64	80.13
	Max.	7.07	0.18	0.00	12.57	80.25
" " Trémosna: Bohemia	Min.	13.78	4.69	0.16	1.10	80.27
	Max.	7.94	0.03	0.09	11.03	80.91
" " Hausham: Bavaria.	Min.	10.48	0.07	0.19	9.28	79.98
	Max.	5.71	0.14	0.08	14.86	79.21
" " Miesbach: Bavaria.	Min.	11.46	0.07	0.07	8.66	79.74
	Max.	5.42	0.03	0.02	15.00	79.53
" " Bohemia............	Min.	17.48	1.21	0.06	3.13	78.12
	Max.	12.20	?	0.30	7.87	?
" " the Ruhr: General	Min.	16.45	1.94	1.45	1.52	78.64
Erbstolln........	Max.	3.95	0.06	0.00	16.41	79.58
" " the Ruhr: Gelsen-	Min.	10.46	0.11	0.11	8.58	80.74
kirchen..........	Max.	5.44	0.12	0.10	14.15	80.19
" " Saarbruck: Saint-	Min.	10.73	0.15	0.30	7.36	81.46
Ingbert............	Max.	7.48	0.07	0.10	11.91	80.44
" " Saarbruck: Mittel-	Min.	13.30	0.61	0.33	4.13	81.63
bexbach........	Max.	8.44	0.19	0.16	10.58	80.63
" " Saarbruck: Heinitz	Min.	14.62	2.07	1.00	2.07	80.24
	Max.	6.49	0.07	0.06	12.70	80.68
" " Saarbruck: mixed ..	Min.	10.22	0.22	0.07	8.57	80.92
	Max.	8.21	0.04	0.02	10.64	81.09
" " Bohemia............	Min.	15.50	0.74	0.33	1.67	81.66
	Max.	8.48	0.08	0.07	9.69	81.68
" " "	Min.	9.61	0.16	0.08	9.47	80.68
	Max.	7.00	0.11	0.05	12.70	80.14
" " Saxony............	Min.	13.80	0.33	0.30	4.36	81.21
	Max.	7.60	0.16	0.09	11.53	80.62
" " Silesia.............	Min.	11.4	0.15	0.04	7.45	81.22
	Max.	8.07	0.10	0.09	10.73	81.01
" " Bavaria: Peissen-	Min.	13.96	1.46	0.79	2.93	80.86
berg............	Max.	7.85	0.07	0.13	10.57	81.38
Lignite from Bohemia...... ..	Min.	14.91	1.04	0.60	2.92	80.53
	Max.	6.36	0.16	0.23	13.15	80.10
Coke from Saarbruck.........	Min.	14.87	0.13	0.09	4.16	80.75
	Max.	8.01	0.03	0.00	10.87	81.09

The data in the above table show that when air to the amount of 15 cubic metres and over per kilogram (200 cubic feet per pound) is used, corresponding to a maximum of 14 per cent of carbonic acid in the waste gases, the loss in hydrogen is very small. With 12 per cent of carbonic acid the hydrogen loss amounts to only a few thousandths.

CALCULATION OF THE VOLUME FROM ANALYSIS.

To calculate this volume, determine the weight of carbon in a unit of volume, and knowing the weight of carbon furnished by the coal, determine the volume corresponding to the unit of weight. The unit of volume for the gas is the cubic metre, and the unit of weight, the kilogram.

Carbon exists in the waste gases as carbonic acid, carbonic oxide, and hydrocarbons; when we do not know the composition of the hydrocarbons, we consider the carbon and hydrogen as free, and that the carbon is in the state of vapor.

To determine the weight of carbon contained in these different gases, reduce their volumes to kilograms, and by means of their molecular (or equivalent) weights and that of carbon make the calculation.

1 litre of CO_2 at 0° and 760 mm. weighs 1.966 grams.
1 "   " CO "   "   "   "   "   "   1.251 "
1 "   " C vapor   "   "   "   1.072 "

Molecular weight of carbon.............. 12
     "   "   " CO_2................ 44
     "   "   " CO.................... 28

The weight of a volume v of carbonic acid is $v \times 1.966$, and as 44 of carbonic acid contain 12 of carbon, then the weight of carbon would be as 44 : 12 or as 11 : 3. Then

$$\frac{v \times 1.966 \times 3}{11} = 0.536v.$$

CALCULATION OF THE VOLUME FROM ANALYSIS.

The weight of carbonic oxide of volume v' is $1.251v'$, and as 28 of carbonic oxide contains 12 of carbon, the ratio becomes $28:12 = 7:3$. We then have

$$\frac{v' \times 1.251 \times 3}{7} = 0.536v'.$$

The weight of a volume of carbon vapor is $v'' \times 1.072$.

To calculate the weight of carbon in a cubic metre of gas, multiply the added volumes of CO_2 and CO by the coefficient 0.536. Multiply the volume of carbon vapor by 1.072, and add this product to that obtained above. The sum is the weight of carbon per cubic metre,

$$C = 0.536(v + v') + 1.072v''.$$

If the gas contains, per cubic metre, 60 litres of carbonic acid, 10 of carbonic oxide, and 1 of carbon vapor, we will have

$$c = 0.536(60 + 10) + 1.072 \times 1 = 38.592 \text{ grams carbon.}$$

From the ratio of carbon of the coal consumed and that in the gas the volume of combustion gases is deduced.

To calculate this, subtract the carbon of the cinders from that of the original coal. If the coal contains 81 per cent carbon and leaves 6 per cent of cinders containing 10 per cent of carbon, then the amount of carbon burnt will be

$$81 - (0.10 \times 6.0) = 81 - 0.6 = 80.4.$$

We then have

$$38.592 : 1000 = 804 : 20.830 \text{ litres.}$$

A kilogram of coal produces, then, 20.83 cubic metres of gas at 0° and 760 mm.

The general formula is

$$V = \frac{C - c}{(v + v')0.536 + 1.072v''},$$

in which

V = volume of waste gases at 0° and 760 mm. in cubic metres;
v = " " CO_2 in litres per cubic metre of gases;
v' = " " CO " " " " " "
v'' = " " carbon vapor per cubic metre of gases;
C = weight of carbon in grams, contained in 1 kilogram of coal;
c = weight of carbon in grams, contained in cinders from 1 kilogram of coal.

NOTE.—The above calculation in English units would be as follows:

Weight of 1 cubic foot of carbonic acid............ 0.12274 lb.
" " 1 " " " " oxide............ 0.07811 "
" " 1 " " " " carbon vapor............ 0.06693 "

$$\frac{v \times 0.12274 \times 3}{11} = 0.0335v.$$

$$\frac{v' \times 0.07811 \times 3}{7} = 0.0335v'.$$

$0.06693v''$ = weight of carbon in vapor.

$C = 0.0335(v + v') + 0.06693v''.$

1000 cubic feet of gases having 60 cubic feet of CO_2, 10 cubic feet of CO and 1 cubic foot of C vapor would give

$C = 0.0335(60 + 10) + 0.06693 \times 1 = 2.412$ lbs. carbon.

1 pound of coal has 80.4 per cent carbon; then

2.412 : 1000 = 0.804 : 333⅓ cubic feet of gases produced from 1 lb. of coal.

The general formula is

$$V = \frac{C - c}{0.0335(v + v') + 0.06693v'''}$$

in which

V = volume in cubic feet of gases produced;
v = " of CO_2 in cubic feet per 1000 cubic feet;
v' = " " CO " " " " "
v'' = " " carbon vapor in cubic feet per 1000 cubic feet;
C = weight of carbon in coal in thousandths of a pound;
c = " " " " cinders per pound of coal in thousandths.

CALCULATION OF VOLUME OF AIR SUPPLIED.

The volume of combustion-gases just determined is less than that of the air supplied. Oxygen in forming carbonic acid produces a volume equal to itself; hence there is no change.

$$C + \underset{\text{2 vols.}}{O_2} = \underset{\text{2 vols.}}{CO_2}$$

Oxygen in forming carbonic oxide produces twice the volume.

$$C + \underset{\text{1 vol.}}{O} = \underset{\text{2 vols.}}{CO}$$

Hence there is an increase in volume.

Carbon vapor and hydrogen as free gases or as hydrocarbons increase the volume but slightly. In forming sulphurous acid with sulphur there is no change of volume.

$$S + \underset{\text{2 vols.}}{O_2} = \underset{\text{2 vols.}}{SO_2}$$

Another slight cause of increase is setting free the nitrogen of the coal; but this is inappreciable. 1 per cent of nitrogen forms only 0.1 per cent of the entire volume of gases formed.

It might be said that, excepting the oxygen changing to water and disappearing by condensation, all the modifications of gaseous volume may be neglected, the increase being more than compensated by the loss due to oxygen. This elimination of oxygen must be allowed for, however.

A coal containing 4 per cent of hydrogen requires eight times such weight to form water, or 40 grams of hydrogen need 320 grams of oxygen. 1 litre of oxygen weighs 1.430 grams, then 320 grams measure $\frac{320}{1.430} = 223.7$ litres (7.9 cubic feet). (Or 1 lb. of such coal would need 3.6 cubic feet of oxygen.)

These 223 litres must be added to the volume of the waste gases produced by the coal to obtain the original volume of air introduced. A coal containing 5 per cent of hydrogen would use 279 litres.

The volume of oxygen needed for various percentages of hydrogen is as follows:

				Per kilo of coal.	Per lb. of coal.
1% hydrogen	uses of	oxygen		55.9 litres,	0.9 cubic feet.
2	"	"	"	112 "	1.8 " "
3	"	"	"	168 "	2.7 " "
4	"	"	"	223 "	3.6 " "
5	"	"	"	279 "	4.5 " "

Calling H the per cent of hydrogen, the formula given above becomes

$$V' = \frac{C-c'}{(v+v')0.563 + 1.071v''} + 55.9\,H,$$

or

$$V' = \frac{C-c'}{0.0335(v+v') + 0.06693v''} + 0.9\,H.$$

To make this applicable to normal air saturated with moisture at 0° C. and 760 mm. (32° F. and 29.922 inches) containing 0.4 per cent of CO_2, we must divide by 99.12, the composition of air being:

Nitrogen....................................	78.35
Oxygen..	20.77
Water................................. 0.84	
Carbonic acid.........,................ 0.04	0.88
	100.00

And $100 - 0.88 = 99.12$. The formula then becomes

$$V'' = \frac{C-c'}{(v+v')0.567 + 1.0806v''} + 55.9\,H,$$

or

$$V'' = \frac{C-c'}{0.0337(v+v') + 0.06752v''} + 0.9\,H.$$

CALCULATION OF WEIGHT OF WASTE GASES FROM ANALYSIS.*

Two methods of calculating from the analysis by volume of the dry chimney gases the number of pounds of dry chimney gases per pound of carbon, or the weight of air supplied per pound of carbon, have been given by different writers. These may be expressed in the shape of formulæ as follows:

(A) Pounds dry gas per pound $C = \dfrac{11CO_2 + 8O + 7(O + N)}{3(CO_2 + CO)}$;

(B) Pounds air per pound $C = 5.8 \dfrac{2(CO_2 + O) + CO}{CO_2 + CO}$.

Formula A may be derived from the method of computation given in Mr. R. S. Hale's paper on "Flue Gas Analyses," *Transactions A. S. M. E.*, vol. XVIII. p. 901, and formula B from the method given in Peabody and Miller's *Treatise on Steam-boilers*. Both are based on the principle that the density, relatively to hydrogen, of an elementary gas (O and N) is proportional to its atomic weight, and that of a compound gas (CO and CO_2) to one half its molecular weight. Both formulæ are very nearly accurate when pure carbon is the fuel burned; but formula B is inaccurate when the fuel contains hydrogen, for the reason that that portion of the oxygen of the air-supply which is required to burn the hydrogen is contained in the chimney gas as H_2O, and does not appear in the analysis of the dry gas.

The following calculations of a supposed case of combustion of hydrogenous fuel illustrates the accuracy of formula A and the inaccuracy of formula B: Assume that the coal has the following analysis: C, 66.50; H, 4.55; O, 8.40; N, 1.00; water, 10.00; ash and sulphur, 9.55; total, 100. Assume

---

* William Kent in Report of Committee on Boiler-tests, A. S. M. E., 1897.

also that one tenth of the C is burned to CO, and nine tenths to CO_2; that the air supply is 20 per cent in excess of that required for this combustion; that the air contains one per cent by weight of moisture; and that the S in the coal may be considered as part of the ash. We then have the following synthesis of results of the combustion of 100 pounds of coal:

		O from Air.	N = O × 11/3.	Total Air.	CO_2	CO	H_2O
59.85 lbs.	C to CO_2 × 2⅔	159.60	534.31	693.91	219.45	.....	.....
6.65 "	C to CO × 1⅓	8.87	29.70	38.57	......	15.52	.....
3.50 "	H to H_2O × 8	28.00	93.74	121.74	......	......	31.50
		196.47	657.75	854.22	......	......	.....
1.05 "	H to H_2O ⎫	......	......	......	......	......	9.45
8.40 "	H to H_2O ⎭						
10.00 "	Water	......	......	......	......	......	10.00
1.00 "	N	......	1.00	......	......	......	......
9.55 "	Ash and S	......	......	......	......	......	... .
100.00		......	......	......	......	......	......
Excess of air 20 per cent.		39.29	131.55	170.84	......	......	.....
		......	......	1025.06	......	......	.....
Moisture in air 1 per cent.		......	......	......	......	......	10.25
Total wt. of gases, 1125.67 =		39.29	790.30	......	219.45	15.52	61.20
Total dry gases, 1064.56							
		O	N		CO_2	CO	
Total dry gases, by weight, %		3.69	74.24	......	20.61	1.546	.....
Total dry gases, by volume, %		3.508	80.656	......	14.252	1.584	.....

Total gases 1125.76 + ash and S 9.55 = 1135.31 total products.
Total air 1025.06 + moisture in air 10.25 + coal 100 = 1135.31.
Dry gas per pound coal 10.6456; per pound carbon = 10.6456 ÷ 665 = 16.008.
Dry air per pound coal 10.2506; per pound carbon = 10.2506 ÷ 665 = 15.414.
Computation of the weight of dry gas and of air per pound C:
Formula A:

$$\text{Dry gas per pound C} = \frac{14.252 \times 11 + 3.508 \times 8 + 82.240 \times 7}{3(14.252 + 1.584)} = 16.008 \text{ pounds.}$$

Formula B:

$$\text{Air per pound C} = 5.8 \frac{2(14.252 + 3.508) + 1.584}{14.252 + 1.584} = 13.589 \text{ pounds.}$$

The error in the last result is 15.414 − 13.589 = 1.825 pounds.

Prof. Jacobus recommends the use of the formula

$$\text{Pounds of air per pound C} = \frac{7N}{3(CO_2 + CO)} \div 0.77;$$

and in the case given above, where the actual quantity used was 15.414 per cent, his calculated one is 15.434 per cent,—practically the same, and as near as errors of analysis would allow a calculated result.

VOLUME OF WASTE GASES BY THE ANEMOMETER.

The *fan-wheel anemometer* is an instrument to measure the force or rapidity of a current of gas. It consists of a fan-wheel rotated by the moving gas, and which transmits such motion to an index showing the number of revolutions. Burnat used this apparatus to measure the quantity of air passing in under the grate of steam-boilers.

The coefficient to be used in calculating the flow is different for each machine, and must be determined by actual experiment. Burnat's formula,

$$v = 0.120 + 0.130n,$$

means that the velocity is found by multiplying the number of revolutions per second by 0.130 and adding 0.120 to the product.

To obtain satisfactory results with the anemometer, it must be placed in the axis of a perfect cylinder at the depth of a metre, as the indications vary with the position in the flue. The formula needs correction for temperature, but the correction of the apparatus much exceeds this. Burnat compared his results with those obtained from a formula depending on the depression if under the grate (see page 147), and found differences of not more than 5 per cent.

FLETCHER'S ANEMOMETER.

Fletcher's anemometer (Fig. 33) is used in England to ascertain the speed of flow in chimneys and flues. In its simplified form it is quite serviceable. It is based on the movement of a column of ether in a U-tube.

The ends of the glass tubes a, b are placed in the flue a little less than one sixth of its diameter. The straight end a

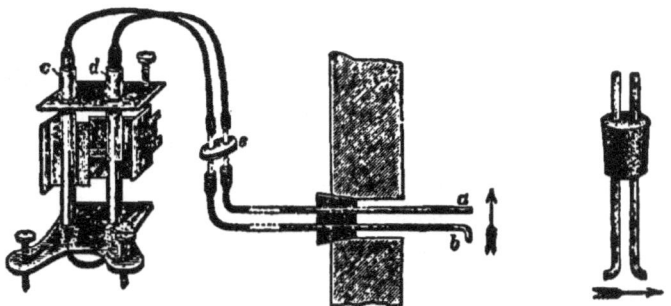

FIG. 33.—FLETCHER'S ANEMOMETER.

should be parallel to the direction of the current, the end b being at right angles to this. Hunter proposed bending both ends in opposite directions, to obviate the error caused if the tubes were not so placed. These tubes communicate with the ether tube cd. The draught across the tubes causes the ether to rise in a by aspiration and to fall in b by pressure. The difference of level is read, and then the tubes are turned around 180°, so as to reverse their positions, and the difference of level read again. The sum of the two differences is called the anemometer reading, and by means of tables the velocity of the current is ascertained.

The same trouble is common to all anemometer methods. The flue feeding the fire receives only the air passing in under the grate. Whatever passes in by the doors or through cracks escapes accounting. On account of this it is certain that the calculations based on anemometer readings are lower than th   al air supply.

SEGUR'S DIFFERENTIAL GAUGE.

This gauge (Fig. 34) consists of a U-tube of $\frac{1}{2}$-inch glass, surmounted by two chambers of $2\frac{1}{2}$ inches diameter. Two non-miscible liquids of different colors, usually alcohol and paraffin oil, are put into the two arms, one occupying the portion AB, the other the portion BCD. The movement of the line of demarcation is proportional to the difference in area of the chambers and the tube adjoining. A movement of 2 inches in the column represents $\frac{1}{4}$-inch difference pressure or draft.

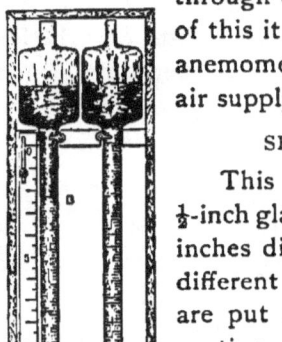

FIG. 34.
SEGUR GAUGE.

HIRN'S METHOD.

The apparatus used by Burnat as a check on his own calculations was devised by Hirn, and is based on the formula of the rate of flow of compressed gases from a reservoir, friction being neglected. The coefficient of reduction used is 0.9, the one given by Dubuisson in his treatise on hydraulics.

The main difficulty consists in measuring the difference of pressure of the atmosphere in the ash pit and that outside, for the depression in the flues in some cases does not exceed a few millimetres of water. Hirn's apparatus removes this difficulty.

Burnat describes it as follows:

When making a test the doors of the ash pit are removed and replaced by a piece of sheet iron, A (Fig. 35), which completely shuts out all access of air except through the opening in the middle, to which is fitted the pipe CD, 13.8 inches diameter and 59 inches long. A tube leads from the front to the apparatus E, devised by Hirn, placed on a table or against the boiler-wall. This apparatus consists of a little gas holder whose upper surface is just one decimeter (3.9

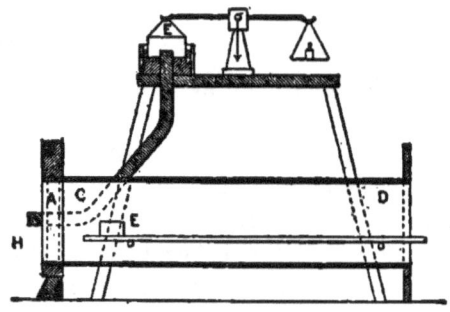

FIG. 35.

inches) on a side. Inside this and above the water level the tube A opens. The bell dips into a vessel of water and is suspended from a balance arm.

The balance being in equilibrium when the atmospheric pressure acts on both sides of the bell, if the interior is connected with the ash-pit the weight needed to restore equilibrium will give a measure of the difference in pressure. The weight of half a gram (7.7 grains) represents one-twentieth millimetre (0.002 inch) of water.

The formula adopted by Hirn is

$$V = S \times 0.9 \sqrt{2g \frac{h \times 0.76(1 + 0.0037t)}{0.0013B}},$$

in which

V = volume of air introduced under the grate in cubic metres;

S = section in square metre of pipe-opening leading air to the ash-pit;

0.9 = coefficient of reduction;

h = difference of pressure expressed in height of water;
B = barometric pressure in the room;
t = temperature of the room;
g = acceleration of gravity = 9.8088 metres.

VOLUME BY AUTOMATIC APPARATUS.

DASYMETER.

Siegert and Durr[*] devised an apparatus called the Dasymeter, which has been introduced in several large works in Europe, where it gives satisfaction.

It consists of a balance enclosed in a cast-iron box with a glass side (Fig. 36). At one end of the beam is a very

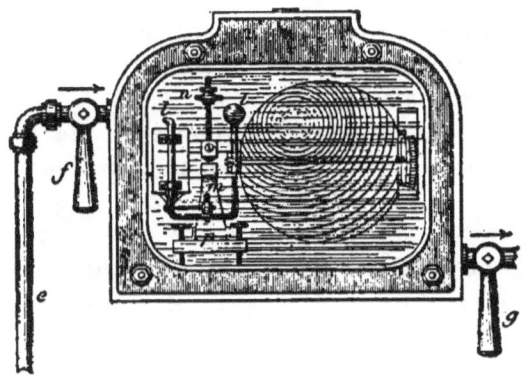

FIG. 36.—DASYMETER.

light glass balloon holding 2 to 3 litres, sealed by fusion. The other end carries a weight balancing the balloon. This weight is formed of a U-tube, ll, containing mercury, and is open at one end; the other end is expanded into a bulb containing air, which is submitted to the variations of pressure and temperature through the mercury. If the pressure of the air increases or diminishes, the mercury rises or falls, and increases or diminishes the weight on the lever. Suppose an

---

[*] Oesterreichische Zeitschrift für B.- und H.-Wesen, XVI. p. 291.

increase of pressure and a lowering of temperature which would diminish the density of the air one half. A corresponding quantity of mercury passes into the arm of the tube, and the original compensating weight is diminished by that amount. A graduated index shows the variations of weight, and hence the variations of density in the gases. An ingenious arrangement allows regulation by rotating the U-tube on the axis pn. The tube is turned slowly around till adjusted, thus changing the length of the lever-arm.

A difference of 1 per cent of carbonic acid causes a difference in weight of 20 milligrams. One litre of air at 0° and 760 millimetres weighs 1294 milligrams; 1 litre of carbonic acid weighs 1967 milligrams; the difference is 673 milligrams. If the gas contains 1 per cent of CO_2, each litre increases 6.73 milligrams in weight; and as the balloon contains 3 litres, it supports an external pressure of more than $3 \times 6.73 = 20.19$ milligrams (0.311 grains).

To prevent action of sulphurous acid the bearings are made of sapphire, onyx, bloodstone, etc., and metallic parts of phosphor-bronze.

To set up the dasymeter, connect pipe e with the boiler-flue before the damper; the tube g leads to the chimney. By this means a current of gas passes through the box, and shows at any time the percentage of carbonic acid. Siegert gives the following results obtained with it, and the corresponding results by analysis:

CO_2 { Dasymeter, 13.0, 13.0, 12.0, 6.25, 2.2, 16.3, 7.5, 12.5
Analysis, 13.0, 12.7, 12.2, 6.00, 2.0, 16.0, 8.0, 13.0

ECONOMETER.

H. Arndt has invented what he calls the "Econometer" (Fig. 37), which is on a similar principle.[*] It consists of a tight cast-iron shell, NN, containing a gas-balance. A pipe,

---

[*] Zeitschrift des Vereines Deutscher Ingenieure, xxxvii. p. 801.

v', 0.4 inch in diameter leads to the inside of the flue before the damper; a second pipe, v'', communicates with the interior of the same flue beyond the damper. In the interior, the tube i' is connected to the upright pipe f, which leads the gas to bell e', and the tube i' to the tubulure g. i' and i'' are of rubber.

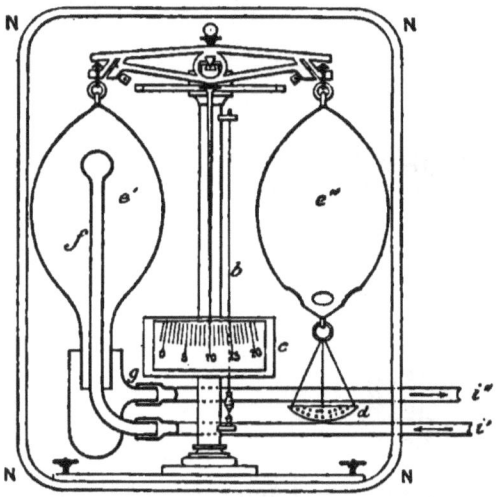

FIG. 37.—ECONOMETER.

The balance is very sensitive, the beam carrying at one end the gas-holder e', open below and containing about 30 cubic inches, and at the other end a second holder of similar size and weight as the first. Attached to the bottom of this one is a pan to hold the balancing weights.

The tube f conducts the gas to the balloon e', which, open below, is freely movable in the cylinder g, by which it produces suction in the tube i'''.

Carbonic acid being heavier than common air (1.96 to 1.29) as well as the other associated gases, it follows that the density of the gases passing through the tubes depends on the carbonic acid content. The scale is divided so that each division shows one per cent of CO_2 in the gases.

GAS-COMPOSIMETER.

The gas-composimeter of Uehling is an apparatus for automatically and continuously determining the quantity of carbonic acid contained in waste gases.

It is based on the laws governing the flow of gas through small apertures.

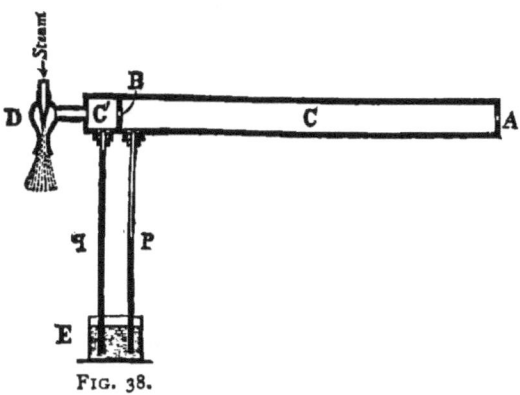

FIG. 38.

If two such apertures, A and B (Fig. 38), form respectively the inlet and outlet openings of chamber C, and a uniform suction is maintained in the chamber C' by the aspirator D, the action will be as follows:

Gas will be drawn through the aperture B into the chamber C', creating suction in chamber C, which in turn causes gas to flow through the aperture A. The velocity with which the gas enters through A depends on the suction in the chamber C, and the velocity at which it flows out through B depends upon the excess of the suction in chamber C' over that existing in chamber C, that is, the effective suction in C'. As the suction in C increases, the effective suction must decrease, and hence the velocity of the gas entering at A increases, while the velocity of the gas passing out through B decreases, until the same quantity of gas enters at A as passes out at B. As soon as this occurs no further change of suction takes place in the chamber C, providing the gas entering at A and passing out at B be maintained at the same temperature.

If from the constant stream of gas, while flowing through chamber C, one of its constituents is continuously removed by absorption, a reduction of volume will take place in chamber C and cause an increase in suction, and consequently a decrease in the effective suction in C'. Hence the velocity of the gas through A will increase, and the velocity through B will decrease, until the same quantity of gas enters at A as is absorbed by the reagent, plus that which passes out at aperture B.

Thus every change in the volume of the constituents we are absorbing from the gas causes a corresponding change of suction in the chamber C.

The apparatus is connected with a regulator, a manometer, and automatic recording register.

TEMPERATURE OF THE WASTE GASES.

As in analyzing coal, cinders, and gases we must have average samples, so in treating of waste gases we need average temperatures. It is not enough to take the temperature occasionally with the thermometer; it varies too much from time to time, even if the readings are taken frequently. We must have some method of obtaining the average temperature of the gas current, and this can be accomplished by means of a heat reservoir introduced into the flue.

For this purpose one was devised by Scheurer-Kestner of a type which has been repeatedly copied and modified. It consists of an iron tube, bb (Fig. 39), placed in the flue so that the upper end, covered with an insulating material, is let into the wall to about one half its thickness, the remainder hanging free in the flue. This tube is filled with paraffin, and in this is inserted the thermometer. The large mass of the paraffin is acted on by the mean temperature, but is uninfluenced by any slight momentary changes which may occur. A self-registering thermometer is very advantageous, but readings at intervals of half an hour are sufficient ordinarily. Of course the opening around the tube should be packed so as to prevent all possible ingress of cold external air.

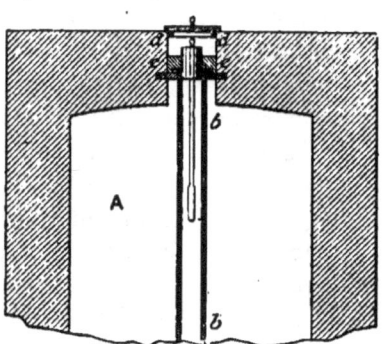

FIG. 39.—FLUE THERMOMETER.

Occasionally mercury is used instead of paraffin. This renders the average of the heat more exactly, perhaps, but has the disadvantage of being much heavier and much more expensive. There are also many difficulties in handling it which do not obtain with paraffin. The paraffin should be well refined, and have a high melting-point.

THE PNEUMATIC PYROMETER.

Uehling's pneumatic pyrometer is based on a principle analogous to that of the gas-composimeter, and is now in use in many places, automatically measuring the temperatures of chimneys and furnaces for all temperatures up to 3000° F., and registering the same on cards. The apparatus has been tested at the Stevens Institute of Technology, and the indications pronounced reliable. It cannot be safely used continuously for temperatures above 2500°, but at that temperature and lower it works well and satisfactorily for months without requiring any readjustment. The automatic register is very sensitive, and can be easily adjusted for a new range of temperatures at any time.

An explanation of the principle of its working is given in the inventor's own words:

"*The Pneumatic Pyrometer* is based on the laws governing the flow of air through small apertures.

"If two such apertures A and B (Fig. 38) respectively form the inlet and outlet openings of a chamber C, and a uniform suction is created in the chamber C' by the aspirator D, the action will be as follows:

"Air will be drawn through the aperture B into the chamber C', creating suction in chamber C, which in turn causes air from the atmosphere to flow in through the aperture A. The velocity with which the air enters through A depends on the suction in the chamber C, and the velocity at which it flows out through B depends upon the excess of suction in C' over that existing in the chamber C, that is, *the effective suction* in C'. As the suction in C increases, the effective suction must decrease, and hence the velocity at which air flows in through the aperture A increases, and the velocity at which air flows out through the aperture B decreases, until the same quantity of air enters at A as passes out at B. As soon as this occurs no further change of suction can take place in the chamber C.

"Air is very materially expanded by heat. Therefore the higher the temperature of the air the greater the volume, and the smaller will be the quantity of air drawn through a given aperture by the same suction. Now if the air as it passes through the aperture A is heated, but again cooled to a lower fixed temperature before it passes through the aperture B, less air will enter through the aperture A than is drawn out through the aperture B. Hence the suction in C

must increase and the effective suction in C' must decrease, and in consequence the velocity of the air through A will increase and the velocity of the air through B will decrease, until the same quantity of air again flows through both apertures. Thus every change of temperature in the air entering through the aperture A will cause a corresponding change of suction in the chamber C. If two manometer-tubes p and q, Fig. 38, communicate respectively with the chambers C and C', the column in tube q will indicate the constant suction in C' and the column in tube p will indicate the suction in the chamber C, which suction is a true measure of the temperature of the air entering through the aperture A.

DETERMINATION OF THE CARBON IN SMOKE.

SOOT or black forms from quick cooling of the hydrocarbons, temporarily dissociated by high temperatures. Fuels having no hydrogen as hydrocarbons, never produce smoke; pure charcoal, coke, or graphite never smokes. Soft coal, on the contrary, produces more as the air-supply grows less.

Sainte-Claire Deville proved that a compound gas when heated sufficiently separates into its elements; a sudden cooling now will give a simple mixture instead of the original combination. A slow cooling, however, reproduces the original gas. Berthelot proved, on the other hand, that new compounds are formed on heating the hydrocarbons to high temperatures, a part of the carbon being deposited as soot. These two phenomena undoubtedly go on together in smoke production.*

If a metal tube be put in the gas current over a grate at a short distance from the fire, the hottest gases will be col-

---

* Bunte gives some analyses of smoke-black:

	C	H
1	97.2	2.8
2	97.3	2.7
3	98.5	1.5

lected. Pass a stream of cold water through a pipe in this gas-current and a large quantity of black will be deposited. On stopping the water flow and inclining the tube a little the carbon disappears gradually, and when the temperature of the tube attains that of the gas, no black will be deposited. Cool it again, and more black forms immediately.

Combustion gases meet with surfaces relatively cold in the boiler sides or flues, or even in colder currents of gas or air passing in through the grate. This produces a quick cooling, and consequent formation of black.

Experiments made at Mulhouse in 1859 by Burnat showed an advantage gained in steaming by producing smoke, rather than introducing too great excess of air. The experiments showed that the loss in carbon was quite small, and these results have been confirmed by others since. E. R. Tatlock of Glasgow finds 60 per cent combustible matter in soot, and obtained 51.46 grains per cubic foot of furnace gases.

To determine the amount of carbon in smoke, Scheurer-Kestner used a glass organic analysis apparatus, the tube having in the middle loosely packed asbestos for about 8 inches, which was kept in place by platinum spirals. One end was drawn out to connect with the absorption apparatus, and the other end placed in the flue. After igniting and cooling the asbestos the small end is connected with an aspirator and the gas drawn slowly through. The carbon is all stopped by the asbestos, which becomes black for a short distance. When sufficiently collected, dry the tube at 100° C., heat to redness, and pass a stream of oxygen through it, collecting the carbonic acid formed.

As an example Scheurer-Kestner gives the following:

Waste gases, reduced to 0° and 760 mm.   86 litres.
Time of sampling.....................   1 hour.

Composition of gas:

CO_2............................ 8.5 per cent.
Excess of air................ 53.4
Nitrogen and residue......... 38.1
CO_2 from the combustion.............. 0.070 gram.
Equivalent to carbon.................... 0.019 "

By the analysis of the gases and that of the coal the quantity of air consumed was calculated. Knowing the volume of air used for the coal, its composition, and the proportion of carbon as black in the gases, the loss due to such formation was calculated.

Kind of Coal.	Waste Gases per Pound of Coal.	Black.	
		Per Cubic Foot of Gas.	Per Cent Calories of Heat of Combustion.
	cubic feet.	grains.	
Ruhr...............	135	15.43	1.1
"	143	7.41	0.6
"	169	0.72	0.07
"	184	6.74	0.2
"	189	1.19	0.1
"	205	2.03	0.1
Hausham............	163	20.49	2.1
"	217	6.79	0.8
"	233	5.71	0.7
"	278	6.48	1.0
"	293	3.70	0.6
Miesbach............	129	1.08	0.1
"	155	6.64	0.8

Under the most unfavorable conditions for feeding the air, the loss due to formation of black does not exceed 2 per cent, even with smoky coal. Ronchamp coal gave the following results:

Feeding 240 cubic feet of air per pound of coal gave a gas containing 8.5 per cent of carbonic acid, excess of air 53 per cent, and loss of carbon as black 0.485 per cent.

Feeding 112 cubic feet of air per pound of coal gave a

gas containing 14.8 per cent carbonic acid, 6.7 per cent excess of air, and 0.96 per cent of black.

Saarbruck coal supplied with 155 cubic feet of air per pound gave a gas having 12.8 per cent of carbonic acid, 28.5 per cent excess of air, and 2.03 per cent of black.

These show that in addition to being a sign of diminution in combustible gases, smoke cannot cause a notable saving in fuel if such saving is accompanied by increased waste gases. The sensible heat of a larger volume compensates easily for the advantages resulting from the more perfect combustion of the carbon.

Bunte publishes the following determinations of black:

Several methods have been devised for approximating to the actual quantity of carbon contained in smoke. One is based on the amount of soot deposited on a given surface placed in the chimney. The soot deposits on the upper surface away from the direct current. After being exposed for a few hours the deposit is brushed off and weighed. Another method is by using smoked glasses of different degrees of opacity and ascertaining what depth of color is necessary to make the smoke invisible. An improvement on this method is now being worked out by one of our manufacturers of optical goods, by means of which the glasses are held in a tube and so arranged as to gradually produce the effect, and in such way that it can be measured.

Another method is that devised by Ringelmann, by means of which the blackness of the smoke is compared with a set of ruled lines, so scaled in width of line and space as to produce six different gradations from smokeless through gray and gray-black to dead black. He recommends the preparation of cards 8 inches square, and have them suspended 50 feet from the observer, at which distance the individual lines become indistinct, and only a general tint is observable. The intensity of the smoke is then compared with the cards and recorded as agreeing with card No. 1, 2, or whatever it may be.

The cards are shown in Fig. 40, reduced in size, the actual lines and spaces being as follows:

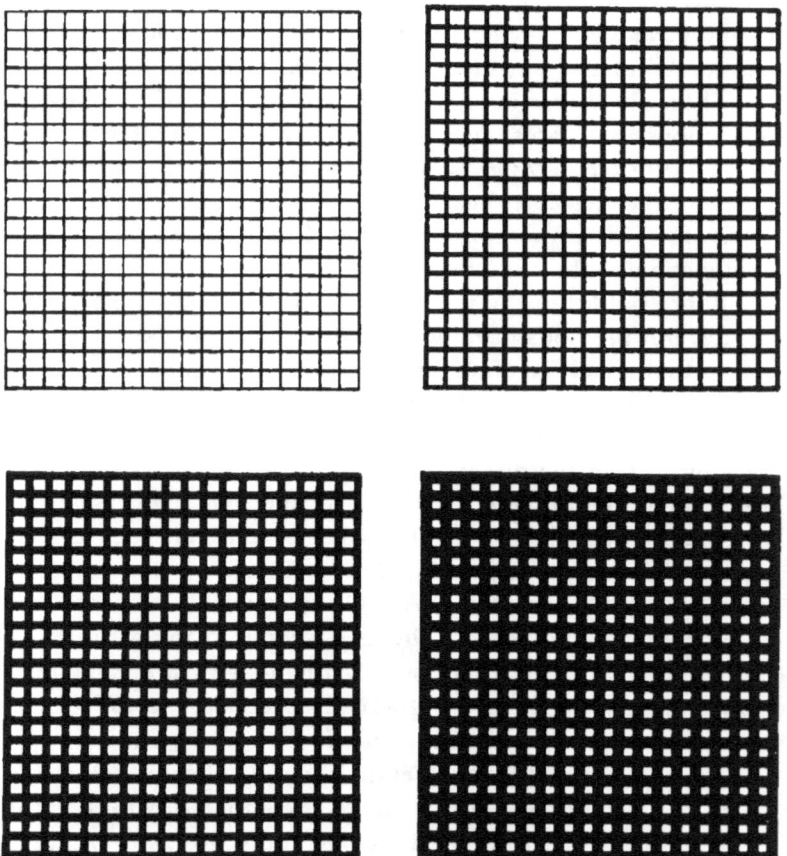

Fig. 40.—Ringelmann Smoke Scale.

Card 0, all white.
Card 1, black lines 1 mm. thick, 10 mm. apart between centres, leaving spaces 9 mm. square.
Card 2, lines 2.3 mm. thick; spaces 7.7 mm. sq.
Card 3, lines 3.7 mm. thick; spaces 6.3 mm. sq.
Card 4, lines 5.5 mm. thick; spaces 4.5 mm. sq.
Card 5, all black.

CHAPTER XII.

CALCULATION OF THE HEAT UNITS.

HEAT OF THE AQUEOUS VAPOR.

The quantity of heat contained in a kilogram or pound of steam at any temperature is

$$Q = 606.5 + 0.305t \text{ calories,}$$
or
$$Q' = 1091.7 + 0.305(t - 32) \text{ B. T. U.,}$$

allowing the specific heat of water to be constant. The number of heat units is considered the same as the temperature.

So that, allowing the average temperature of aqueous vapor to be 150° C., each kilogram at 0° has absorbed a quantity of heat equal to

$$606.5 + 0.305 \times 150 = 652.25 \text{ calories}$$
or one pound has absorbed 1174 B. T. U.

There is a correction to this, since we do not wish the units existing in the steam, but only those added to it from the fuel. We must then deduct that already existing in the water at its entrance to the boiler. If the feed-water be 20° (68° F.) the formula becomes

$$652.25 - 20 = 632.25 \text{ calories,}$$
or
$$1174 - (68 - 32) = 1138 \text{ B. T. U.}$$

HEAT OF WASTE GASES.

The heat carried to the chimney by the waste gases is from several sources:

1. Sensible heat shown by the temperature.
2. Heat of vaporization of the hygroscopic water and the water formed from the hydrogen of the coal.
3. Heat retained by the combustible gases or their heat of combustion.
4. Heat represented by soot or black of the smoke.

1. SENSIBLE HEAT OF THE TEMPERATURE.

The calculation of the water equivalent of the heat carried to the chimney as sensible heat requires the volume, temperature, composition, and specific heat of the constituents.

The specific heats of the usual constituents of waste gases are shown in Table VIII. The specific heats are supposed to be under constant pressure, so as to avoid useless calculations. The hydrocarbons or hydrogen will be omitted for the same reason. Calling v, v', v'', v''' the volumes in cubic metres of the gases nitrogen, carbonic acid, carbonic oxide, and oxygen, we find their respective weights, by multiplying these volumes by the weight per cubic metre,

$$\frac{1.256v}{N}, \quad \frac{1.966v'}{CO_2}, \quad \frac{1.251v''}{CO}, \quad \frac{1.430v'''}{O}.$$

Multiplying these by the specific weights we obtain the value in water,

$$C = 1.256v \times 0.244 + 1.966v' \times 0.217 + 1.251v'' \times 0.245 + 1.430v''' \times 0.217.$$

The equivalent in water c multiplied by the temperature on leaving the boiler gives calories,

$$C = c \times T.$$

A correction of the same kind as that applied to the temperature of the feed-water must be applied. We do not wish the total calories, only those taken up from the coal. From the observed temperature T we must deduct the original temperature t before entering the fire. So that

$$C = c \times (T - t).$$

The general formula then becomes

$$C = [\underbrace{(1.256v)0.244}_{N} + \underbrace{(1.966v')0.217}_{CO_2} + \underbrace{(1.251v'')0.245}_{CO}$$
$$+ \underbrace{(1.430v''')0.217}_{O}] (T-t).$$

As an example, suppose the following composition:

Nitrogen...... 81.25 } = { Air in excess...... 23.04 (4.84 × 4.761)
Oxygen........ 4.84 }   { Nitrogen.......... 63.05 (81.25−4.84−23.04)
Carbonic acid.. 13.08 .... 13.08
Carbonic oxide. 0.83........................... 0.83
                ───   ───
                100.00   100.00

and that the temperature $(T - t)$ is 130°. Then

Nitrogen........ 1.256 × .8125 × 0.244 = 0.249
Carbonic acid.... 1.966 × .1308 × 0.217 = 0.055
Carbonic oxide... 1.251 × .0083 × 0.245 = 0.002
Oxygen.......... 1.430 × .0484 × 0.217 = 0.015
                         ──────   ─────
                         1.0000   0.321

The value in water for 1 cubic metre is 0.321 kilogram, which at 130° give

$$0.321 \times 130 = 41.7 \text{ calories.}$$

If the volume of the gases was 8.938 cubic metres per kilogram of coal, the calories carried to the chimney would be

$$\frac{8.938 \times 41.7}{100} = 372 \text{ calories. (669.6 B. T. U.)}$$

The same result can be reached more quickly by taking the ratio of the specific heats to the volume (Table VIII).

N	.8125 × 0.306 = 0.249
CO_2	.1308 × 0.426 = 0.055
CO	.0083 × 0.306 = 0.002
O	.0484 × 0.310 = 0.015
1.0000	0.321

0.321 × 130 × 8.938 = 372 calories.

This may be still further simplified in practical work with the combustion under normal conditions. Base the calculation on the proportion of carbonic acid, using 0.306 as coefficient for the remaining gases. Then

$$C = (0.426v + 0.306R)(T-t)$$

v CO_2	0.1308 × 0.426 = 0.055
R N, CO, and O	0.8692 × 0.306 = 0.266
	0.321

By means of the coefficients in Table IX we can still further shorten the calculation. By this table we get directly

0.321 × 130 × 8.938 = 372 calories.

The loss of heat due to temperature of the waste gases varies according to the condition of the boiler, its surface for radiation, the grate surface, and the air supply. With the most advantageous cases, and moderate combustion, the gas temperature at the exit does not exceed 150° (302° F.), and the loss, 5 or 6 per cent of the total heat of combustion. It may reach 10 per cent, and in some cases even more.

2. HEAT OF THE HYGROSCOPIC AND COMBUSTION WATER.

During combustion, coal furnishes a quantity of aqueous vapor from its hygroscopic water and its hydrogen; the latter is determined by multiplying the weight of hydrogen by 9. This is added to the hygroscopic water, and the formula

$$(606.5 + 0.305t) - t'$$

applied; t being the temperature of the vapor in the gases (equal to that of the gases), and t' being that of the external air. Besides this, however, we must consider the specific heat of the aqueous vapor, 0.475. Each kilogram still absorbs 0.475 multiplied by the number of degrees of temperature above 100°, and the formula becomes

$$x[(606.5 + 0.305t) - t' + 0.475(t - 100)],$$

x being the quantity of water, in kilograms, furnished by the coal.

Suppose a coal contains 15 grams per kilogram of hygroscopic water and 45 grams of hydrogen, as follows:

Hygroscopic water.	15
Carbon.	735
Hydrogen.	45
Nitrogen and oxygen.	50
Ash.	160
	1000

Hydrogen 45 produces $9 \times 45 = 405$ grams, to which add the 15 grams of hygroscopic water, $405 + 15 = 420$ grams. The heat necessary to vaporize this, increased by that corresponding to the temperature of the gases passing up the chimney, represents the heat lost.

If the flue temperature is $145° = t$, and the external air $17.5° = t'$, we have

$$0.420[(606.5 + 0.305 \times 145) - 17.5 + 0.475(145 - 100)$$
$$= 274.9 (494.8 \text{ B T. U.}).$$

If the heat of combustion of the coal is 7000 calories, then the loss is

$$\frac{274.9}{7000} = 3.92 \text{ per cent.}$$

The loss due to these causes in an average coal (4–5 per cent hydrogen and 1 to 2 per cent moisture) is usually from 2 to 4 per cent.

3. CALORIES OF THE COMBUSTIBLE GASES.

Carbonic oxide is always present in variable quantities, often hydrocarbons and sometimes hydrogen. This refers to ordinary fuel and the usual methods of burning. The quantity of unburnt gases depends on the kind of fireplace used and the system of charging. Thick charges of fuel always increase the volume of unburnt gases; the smallest amount being obtained from small, equivalent charges, fed frequently and using 30 to 50 per cent more air than the theoretical quantity.

To determine this loss we may commence with the volume or the weight corresponding to 1 kilogram of coal burnt. The calculation is given on pages 137 and 138. No account need be made of the temperature, the calculation of loss due this having been made on page 161 for all gases, and therefore for these gases.

The calorific coefficients of the unburnt gases, referred to a cubic metre at 0° and 760 mm. pressure, are

	Weight per cub. m. in Kilograms.	Heat of Combustion.	
		Per Kilo.	Per Cubic Metre.
Hydrogen	0.089	34500	3091
Carbonic oxide	1.251	2435	3043
Methane (CH_4)	0.715	13343	10038
Carbon vapor	1.073	11328	12143

The weight and heat of combustion of carbon vapor are given, as most of the time we do not know the molecular condensation of the hydrocarbons; usually the ultimate composition is all that is known. Hence the hydrogen and carbon must be given their heat values as though free. Fortunately they occur in only small percentages, and the error introduced by so doing is small.

Suppose a gas to analyze

Carbonic oxide	1.0
Carbonic acid	13.0
Methane	1.0
Oxygen	6.0
Nitrogen	79.0
	100.0

Assuming that the air has been fed at the rate of 10 cubic metres per kilogram (160.5 cubic feet per pound), and that the coal has a heat value of 8000 calories (14400 B. T. U.), we will have, for 10 cubic metres,

Carbonic oxide	0.1	cubic metres.
Carbonic acid	1.3	"
Methane	0.1	"
Oxygen	0.6	"
Nitrogen	7.9	"
	10.0	

Then

CH_4, 0.1 cub. m. @ 0.715 = 0.0715 kilogram;
CO, 0.1 " " @ 1.251 = 0.1251 "

and

$0.0715 \times 13343 = 933.7$ calories;
$0.1251 \times 2435 = 305.0$ "

Total............1238.7 "

The loss, then, is 1238.7 in 8000, or 15.48 per cent.

If instead of knowing the proportion of the hydrocarbons we know only that of carbon and hydrogen, the heat values calculate separately. Then, instead of methane 0.1, there would be carbon 0.05, and hydrogen 0.2. Then the calculation would be

$0.2 \times 0.089 = 0.0178;\quad 0.0178 \times 34500 = 614.1$
$0.05 \times 1.073 = 0.0536;\quad 0.0536 \times 8137 = 436.1$
$0.1 \times 1.251 = 0.1251;\quad 0.1251 \times 2435 = 305.0$

$\overline{}$
1355.2 calories

The difference, $1355.2 - 1238.7 = 116.5$ calories, or 0.9 per cent of the calories lost, or $15.48 \times .009 = 0.138$ per cent of the total calories of the coal, which is small compared with other sources of error.

By employing Table VII we may dispense with reducing the volumes to weights, thus:

$$\begin{array}{lrcr}
\text{Hydrogen} \ldots\ldots\ldots & 0.2 m' & \times\ 3091 = & 618 \\
\text{Carbon vapor} \ldots\ldots & 0.05 & \times\ 8722 = & 436 \\
\text{Carbonic oxide} \ldots & 0.1 & \times\ 3043 = & 304 \\
\hline
& & & 1358
\end{array}$$

The preceding is an exaggerated case; as usually, with ordinary working, the loss is from 2 to 7 per cent, rarely exceeding the latter. Either method of calculation may be used, then, without risk of causing an error of importance.

4. CALORIES DUE TO THE SOOT.

The soot in smoke consists of carbon with a trace of hydrogen. It can be calculated as all carbon without appreciable error and with the coefficient 8137. Knowing the volume of gases produced by 1 kilogram and its content in black (page 154), calculate the number of calories. Under the most favorable conditions for smoke production the loss does not exceed 1 per cent, and is generally less than one half that amount.

DISTRIBUTION OF CALORIES—LOSS.

The difference between heat units accounted for and those possible is considered as resulting from radiation by surfaces not available for producing steam. The following is taken from Scheurer-Kestner's results with a three-tube steam boiler followed by a reheater. The first column gives results obtained with Ronchamp coal in 1868, the second results with Nixon's Navigation Co.'s coal in 1881.

	Ronchamp.	Nixon.
Calories in the steam	58 to 67%	74.5%
" " " waste gases	3.8 to 7.7	5.42
" " " unburnt gases	2.4 to 9.7	traces
" " " smoke	0.3 to 0.75	none
" " " aqueous vapor	2.0 to 3.7	2.81
" not accounted for	19.4 to 24.7	17.27

On September 20, 1895, *Engineering* published the results of some experiments made by Bryan Donkin with Nixon's coal on twenty different types of boilers. The following table contains some of them:

Calories.	XII.	VIII.	VI.	VII.	II.	XI.	III.	IV.	XX.	I.
In the steam	78.5	78.3	74.4	71.8	70.4	69.8	67.6	66.2	65.8	63.8
In the waste gases	6.5	14.0	13.8	13.3	13.6	18.0	16.2	22.5	18.0	9.4
In the combustible gases	0.0	1.7	2.4	0.8	0.0	1.2	1.2	0.0	1.6	12.7
Not accounted for	15.0	5.8	9.3	14.0	11.9	10.9	9.6	11.0	14.4	13.9

The calories in the steam	varied from	63.8 to 78.5 per cent.
" " " " waste gases	" "	6.5 to 22.5 " "
" " " " combustible gases	" "	0.0 to 12.7 " "
" " not accounted for	" "	5.8 to 15.0 " "

For the method of properly tabulating the heat balance, see section XXI of the Steam Boiler Code on page 193.

FLAME AND FLAME TEMPERATURES.

Whenever the temperature is sufficiently high to raise a portion of the carbon, hydrogen, or other gaseous combustible to incandescence, flame is produced. The temperature at which this phenomenon occurs varies with the substance burnt. Usually it requires a red heat or higher, but in some cases a much lower temperature suffices: bor-methyl $B(CH_3)_3$ is an example, the flame temperature of which is not high enough to scorch the finger placed in it. It is not necessary that the flame should have solid particles in it, as flame is produced by hydrogen burning under pressure in oxygen; neither is incandescence alone sufficient, as the fire of pure carbon, magnesium, or iron glows but does not flame. Flame is hollow, the combustion occurring on the surface, and this may be easily demonstrated, by drawing off some of the interior unconsumed gases with a tube and burning them.

Bunsen's researches led to the conclusion that the temperature of burning carbonic oxide rapidly rose to 3000° C., and remained stationary till one third of it was consumed; the temperature then fell to 2500° C., at which more burnt; and finally fell to about 1200° C., which temperature was maintained till all the remainder was consumed. Actually the last temperature is soon reached in practice. Berthelot confirms this, but is in doubt whether the loss of temperature is due to dissociation or to change in specific heat. Some hold that part of this loss of heat is caused by its absorption, due to the production of incandescence and its accompanying flame phenomena. A gas raised to incandescence gradually manifests each increment of heat till that point is reached, and beyond this no increase is noticed, all such further increase being consumed by the flame production.

The rate of propagation of flame varies with the pressure and with the material burning. The most rapid rate with coal gas is when it is mixed with 5 parts of air; with marsh gas, 8½ parts of air. It will be noticed that the proportion of oxygen is sensibly less than that required for perfect combustion.

The luminosity depends on the compression of the gases or the air. Hydrogen burning in oxygen at ordinary pressure gives a flame hardly visible at all; with a pressure of 20 atmospheres it becomes quite luminous. Arsenic in burning produces quite a luminous flame at ordinary air pressure; but hardly any in rarefied air. The same is true of carbonic oxide and other gases. The luminosity seems to be in direct proportion to the pressure.

Luminosity seems to be greater with those substances which on burning produce dense vapors. Hydrogen and chlorine produce a vapor twice as heavy as water and the luminosity is much stronger than with the oxygen-hydrogen flame. Carbon and sulphur also produce heavy vapors and much light. Phosphorus burning in oxygen produces the dense heavy phosphoric anhydride and this is accompanied with an almost blinding light.

The length of the flame ordinarily depends on the quantity of hydrogen, and consequently the hydrocarbons contained in, or generated from, the body consumed. With fuels containing high hydrocarbon percentages, flame of almost any desired length can be produced. This is especially the case with gases.

The theoretical temperature of combustion, and hence of the flame, may be calculated by dividing the heat units produced by the specific heats of the products formed. Of course, these theoretical temperatures are never reached in practice, but they serve as aids in determining the value of fuels for certain purposes.

A few typical examples of these calculations will be given.

1. *Hydrogen.* — Hydrogen burnt in oxygen produces 29000 heat units (water considered as vapor); the specific heat of the aqueous vapor produced is 0.475. The hydrogen

uses 8 times its weight of oxygen and generates 9 times the quantity of water.

Then
$$\frac{29000}{9 \times 0.479} = 6727° \text{ C.}$$

Bunsen and Sainte-Claire Deville showed that the highest temperature actually obtained is 2500° C., which may be increased to 2850° C. by a pressure of 10 atmospheres.

The presence of nitrogen modifies the result materially. The quantity of oxygen required, obtained from air, would introduce 26.78 parts of nitrogen, the specific heat of which is 0.244. The equation would then be

$$\frac{29000}{9 \times 0.479 + 26.78 \times 0.244} = 2674° \text{ C.}$$

Bunsen's maximum temperature actually reached was 1800° C.

2. *Carbon.*—Carbon burnt to carbonic oxide consumes 1.33 parts of oxygen, forms 2.33 parts of carbonic oxide, and if burnt in air, introduces 4.46 parts of nitrogen. The specific heat of carbonic oxide is 0.245 and of nitrogen 0.244, as before. The heat units generated are 2435.

For combustion in oxygen the equation would be

$$\frac{2435}{2.33 \times 0.245} = 4265° \text{ C.}$$

In air it would be

$$\frac{2435}{2.33 \times 0.245 + 4.46 \times 0.244} = 1462° \text{ C.}$$

The latter temperature is about the same as that actually observed, and shows that but little dissociation occurs. Owing to the non-volatility of carbon no flame is produced, only an incandescence. The flame we ordinarily see on incandescent carbon is from the burning of carbonic oxide. Carbon burnt to carbon dioxide can be treated similarly; also carbonic oxide burnt to carbon dioxide.

3. *Marsh Gas.*—This gas requires 4 times its weight of oxygen, and produces 2.25 parts of aqueous vapor and 2.75 parts of carbonic acid. If air is used, 13.39 parts of nitrogen are introduced. The heat of combustion is 13343 calories.

The equations are, then,

$$\frac{13343}{2.25 \times 0.479 + 2.75 \times 0.217} = 7971° \text{ C.,}$$

for oxygen and

$$\frac{13343}{2.25 \times 0.479 + 2.75 \times 0.217 + 13.39 \times 0.244} = 2245° \text{ C.,}$$

for combustion in air.

Olefiant gas, acetylene, etc., can be calculated similarly. With a mixed gas, i.e., one containing several gases, account must be taken of each one separately. Producer gas will be given as an example.

4. *Producer Gas.*—The producer gas taken will be assumed to have the following composition by volume:

Carbonic oxide	21.0	per cent.
Hydrogen	11.5	" "
Marsh gas	2.0	" "
Carbonic acid	6.0	" "
Nitrogen	59.5	" "
	100.0	" "

First obtain the weight of the constituents. (See the tables.)

$$0.21 \times 1.2515 = 0.2628$$
$$0.115 \times 0.0896 = 0.0103$$
$$0.02 \times 0.7155 = 0.0143$$
$$0.06 \times 1.9666 = 0.1360$$
$$0.595 \times 1.2561 = 0.7474$$

			CO_2	H_2O	N
CO	0.2628	produces	0.413	....	0.502
H	0.0103	"	....	0.093	0.276
CH_4	0.0143	"	0.039	0.032	0.192
CO_2	0.1360	"	0.136	....	...
N	0.7474	"	....	....	0.747
			0.588	0.125	1.717

Then as the heat of combustion is 747.66 by volume or 874.6 by weight, we have for combustion in oxygen,

$$\frac{874.6}{0.125 \times 0.479 + 0.588 \times 0.217 + 0.747 \times 0.244} = 2350° \text{ C.},$$

and for combustion in air,

$$\frac{874.6}{0.125 \times 0.479 + 0.588 \times 0.217 + 1.717 \times 0.244} = 1112° \text{ C.}$$

5. *Petroleum Oil.*—The oil may be assumed to contain

Carbon.................... 85 per cent.
Hydrogen. .... 15 " "
──────
100

C 0.85 produces .... 3.117 CO_2 and 7.588 N
H 0.15   "   1.35 H_2O .... "   "   4.017 "
                    ──────   ──────   ──────
                    1.35 H_2O 1.117 CO_2 11.605 N

The heat of combustion may be assumed at 10000 calories. Then for combustion in oxygen,

$$\frac{10000}{1.35 \times 0.479 + 3.117 \times 0.217} = 7558° \text{ C.},$$

and for combustion in air,

$$\frac{10000}{1.35 \times 0.479 + 3.117 \times 0.217 + 11.605 \times 0.244} = 2400° \text{ C.}$$

Other oils or solid fuels may be calculated according to this model.

At the end of the volume are given a few of those fuels most commonly used with the theoretical oxygen and air flame temperatures.

WEIGHT AND HEAT UNITS OF CARBON VAPOR.

Two volumes of carbonic oxide are produced from 1 volume of oxygen, and hence from 1 volume of carbon. 1 cubic metre of carbonic oxide weighs 1251 grams. 1 cubic metre of oxygen weighs 1430 grams. 1 cubic metre of carbonic oxide contains, then, one-half a cubic metre of oxygen weighing 715 grams, and one-half a cubic metre of carbon vapor weighing 536 grams. Hence 1 cubic metre of carbon vapor weighs $2 \times 536 = 1072$ grams, and 1 kilogram measures $1 : 1072 = 0.9328$ cubic metre.

Or

1 cubic foot of carbonic oxide weighs 546.78 grains.
1 " " " oxygen weighs....... 624.85 "

One cubic foot CO then contains $\frac{1}{2}$ cubic foot of O and $\frac{1}{2}$ cubic foot of C.

$$546.78 - 312.425 = 234.355,$$

and

$$2 \times 234.355 = 468.71 \text{ grains,}$$

weight of 1 cubic foot of carbon vapor.

One pound of carbon vapor measures 14.93 cubic feet.

If we wish the heat-units of carbon in vapor without the heat of vaporization, multiply the weight of a cubic metre by the heat of combustion of solid carbon. If from wood charcoal,

$$8137 \times 1.072 = 8722 (15699.6 \text{ B. T. U.}).$$

If from diamond,

$$7859 \times 1.072 = 8424 (14963.2 \text{ B. T. U.}).$$

If carbon vapor with its heat of vaporization be wanted, take the heat of combustion of carbonic oxide which contains carbon as vapor and compare it with the heat of combustion of carbon, uniting with the same quantity of oxygen to form

carbonic oxide. In doing so it is supposed that carbon in combining with two atoms of oxygen generates the same quantity of heat with one as with the other, only in the first case part of the heat is used in vaporizing the carbon. This heat is found by subtracting the heat of combustion of the solid carbon from that of the carbon supposed gaseous in carbonic oxide.

One kilogram of carbon unites with 1.333 kilograms of oxygen to form 2.333 kilograms of carbonic oxide. With diamond there is generated 2405 calories. The 2.333 kilograms of carbonic oxide in becoming carbonic acid generates $2.333 \times 2435 = 5680$ calories. Then 1 kilogram of carbon in passing from carbonic oxide to carbonic acid generates 5680 calories. We have seen, on the other hand, that 1 kilogram of diamond carbon generates 2405 calories in becoming carbonic oxide. The difference, then, $5680 - 2405 = 3275$ (5895 B. T. U.) calories, represents the heat of vaporization of diamond carbon. With wood charcoal it becomes $5680 - 2489 = 3191$ (5743.8 B. T. U.).

The heat of combustion will be then $7859 + 3275 = 11134$ calories (20041 B. T. U.) for diamond, and $8137 + 3191 = 11328$ calories (20390 B. T. U.) for wood charcoal.

EVAPORATIVE POWER OF FUEL.

The evaporative power of a fuel represents the number of pounds of water at 212° F. that can be evaporated or converted into steam by one pound of the fuel. Water at that temperature is sufficiently heated to vaporize, but needs an addition of force equivalent to that required for the vaporization. This quantity varies for the pressure of the barometer and the temperature of the water, but for the purposes of calculation is considered to be taken at 30 inches of mercury and 212° F. Experiment has shown the equivalent to be 965.7 heatunits (B. T. U.).

To find the theoretical evaporating power of a fuel, then, divide the number of thermal units it generates on combustion by 965.7. For instance, the heat of combustion of a sample of Illinois coal was determined by Prof. Carpenter to be 13200. Its evaporative power would be

$$\frac{13200}{965.7} = 13.67 \text{ pounds.}$$

This means that under the proper conditions one pound of the coal in question would evaporate 13.67 pounds already heated to 212° F.

But this amount of duty is rarely realized. The boiler may not be well built, the setting may be faulty, and there are numerous other chemical or mechanical conditions which modify the yield. With these no rule can be established; each individual case must be allowed for specially. With ashes and moisture, chemical constituents of the coal, the case is different. A percentage allowance for these will usually suffice.

For instance, in the above coal there was 5.12 per cent of water and 15.2 per cent of ash. Then

$$100 - (15.2 + 5.12) \times 13.67 = 12.23 \text{ pounds.}$$

If deemed necessary, a further correction can be made for the water of the coal, which would reduce the evaporation by its own amount. This correction would become

$$12.23 - 0.05 = 12.18 \text{ pounds}$$

as the quantity which should be evaporated with the coal as analyzed.

The quantity of ash produces an effect on the evaporative power aside from its proportional reduction in combustible. This is due to the fact that where a large percentage of ash occurs, the particles of carbon of the fuel are not burnt completely, owing to being enclosed in the ash and consequently shut off from access of air. This is especially the case with those ashes which are easily fuzed by the heat of the fire. Ashes containing carbonates are much more easily fuzed than those containing phosphates or sulphates. On this account a chemical analysis of the ash is at times quite desirable.

Some difference in evaporation is noticed in using the different sizes of coal, more particularly with the fine sizes. With the proper arrangements for burning fires a good yield is obtained, but with the ordinary grates the yield is much lower.

APPENDIX.

REPORT OF THE COMMITTEE ON THE REVISION OF THE SOCIETY CODE OF 1885, RELATIVE TO A STANDARD METHOD OF CONDUCTING STEAM-BOILER TRIALS.

Presented to the New York meeting of the American Society of Mechanical Engineers, December, 1897, and forming a part of the Transactions, Volume XIX.

To the American Society of Mechanical Engineers.

Gentlemen: The undersigned Committee, to which was submitted the revision of the Society Code of 1885, relative to a standard method of conducting steam-boiler trials, reports as follows:

The former Committee gave a full statement of the principles which governed it in the preparation of the Code of Rules at that time recommended. These principles covered the ground in an admirable manner, so far as the practice of boiler-testing had been perfected, and we are in unanimous accord with the sentiments which the report of that Committee expressed. During the interval of twelve years which has passed, methods and instruments have in some measure changed. Improvements have been made in the instruments for determining the moisture in steam. The throttling and separating form of calorimeters have displaced the barrel and other types of steam calorimeters referred to in the previous report. Attention has been devoted to the determination of the calorific value of coal, and a number of coal calorimeters have been brought out and successfully used for this purpose. It has come to be a practice with many experts to include in the table of results of boiler-tests the percentage of "efficiency," or proportion of the calorific value of the coal which is utilized by the boiler. Specifications and contracts are in some cases drawn up, providing for certain percentages of efficiency instead of a specified evaporation. The analysis of flue-gases is receiving more attention than formerly, not only in our educational institutions, but also in the regular practice of engineers who make a specialty of boiler-testing.

Your Committee submits a revised Code, termed the Code of 1897. It is substantially the same as the 1885 Code, with such amendments as the experience of the last twelve years has shown to be desirable.

It is beyond the province of the Committee to recommend instruments of particular makers for obtaining the quality of the steam, the calorific value of the fuel, or any other data relating to the trial; but following the practice of the former Committee, individual members have submitted their views (with the approval of the full membership) in an "Appendix to the 1897 Code," signed by their initials. In this appendix are included some of the articles from the appendix to the former Code, which are thought to be of especial value.

In the matter of instruments for determining the calorific value of fuel, it seems desirable that the Committee should make a recommendation which is as specific as present knowledge and circumstances will warrant. It is agreed that some form of calorimeter in which the coal is burned in an atmosphere of oxygen gas is to be preferred, and it is generally held that the most perfect apparatus thus far brought out is the Bomb Calorimeter, originally designed by Berthelot, and modified by Mahler and Hempel. Several of these instruments are in use in this country, principally in the laboratories of engineering schools; but the apparatus is complicated and expensive, and it is not probable that many engineers will have the instrument as a part of their equipment for testing boilers. It is recommended, therefore, that samples of the coal used in testing boilers be sent for determinations of their heating value to a testing laboratory provided with one of these instruments, or with some instrument which shall be proven to be equally good.

Besides the amendments to the Code of 1885, concerning the determination of "efficiency" and the use of improved steam calorimeters, directions are given for sampling the coal, for determining the heat of combustion from the chemical analysis of coal, and for working out a heat balance. Rules are laid down for finding the quantity of moisture in coal and for making allowance for it. The tabular form of presenting the results of the test is somewhat changed from that of the Code of 1885, and alterations in the text of that Code are made wherever revision seems desirable.

The Committee approves the conclusions of the Committee of 1885 concerning the standard "unit of evaporation" contained in the following extract from the introduction to the Code of 1885:

"It has gradually come to be the custom to reduce all results to the common standard of weight of water evaporated by the unit weight of fuel, the evaporation being considered to have taken place at mean atmospheric pressure, and at the temperature due that pressure, the feed-water being also assumed to have been supplied at that temperature. This is, in technical language, said to be the 'equivalent evaporation from and at the boiling-point' (212 degrees Fahr.), and has now become so generally incorporated into the science and the practice of steam-engineering that your Committee would simply express their approval of the adoption, and recommend the permanent retention of this 'unit of evaporation,' viz., one pound of water at 212 degrees Fahr. evaporated into steam of the same temperature. This is equivalent to the utilization of 965.7 British thermal units per pound of water so evaporated."

The unit of commercial boiler horse-power adopted by the Committee of 1885 was the same as that used in the reports of the boiler-tests made at the Centennial Exhibition of 1876. The Committee of 1885 reported in favor of this standard in language of which the following is an extract:

"Your Committee, after due consideration, has determined to accept the Centennial standard, and to recommend that in all standard trials the commercial horse-power be taken as an evaporation of 30 pounds of water per hour from a feed-water temperature of 100 degrees Fahr. into steam at 70 pounds gauge-pressure, which shall be considered to be equal to $34\frac{1}{2}$ units of evaporation; that is, to $34\frac{1}{2}$ pounds of water evaporated from a feed-water temperature of 212 degrees Fahr. into steam at the same temperature. This standard is equal to 33,305 thermal units per hour."

The present Committee accepts the same standard, but reverses the order of two clauses in the statement, and slightly modifies them to read as follows:

In all standard trials the commercial horse-power shall be taken as $34\frac{1}{2}$ units of evaporation; that is, $34\frac{1}{2}$ pounds of water evaporated from a feed-water temperature of 212 degrees Fahr. into steam at the same temperature. This standard is equivalent to 33,317 British thermal units per hour. It is also practically equivalent to an evaporation of 30 pounds of water from a feed-water temperature of 100 degrees Fahr. into steam at 70 pounds gauge-pressure.*

---

* According to the tables in Porter's Treatise on the Richards Steam-engine Indicator, an evaporation of 30 pounds of water from 100 degrees Fahr. into steam at 70 pounds pressure is equal to an evaporation of 34.488 pounds from and at 212 degrees; and an evaporation of $34\frac{1}{2}$ pounds from and at 212 degrees Fahr. is equal to 30.010 pounds from 100 degrees Fahr. into steam at 70 pounds pressure.

The "unit of evaporation" being equal to 965.7 thermal units, the commercial horse-power $= 34.5 \times 965.7 = 33.317$ thermal units.

The Committee also indorses the statement of the Committee of 1885 concerning the commercial rating of boilers, changing somewhat its wording, so as to read as follows:

"It is the opinion of this Committee that a boiler rated at any stated horse-power should develop that power when using the best coal ordinarily sold in the market where the boiler is located, fired by an ordinary fireman, with a draft at the smoke-box not exceeding $\frac{3}{8}$ inch of water column; and, further, that the boiler should develop at least one third more than its rated power when operated with the best system of firing and with the full draft available."

Respectfully submitted,

CHAS. E. EMERY,[*]
WM. KENT,
GEO. H. BARRUS,
CHAS. T. PORTER,
ROBERT H. THURSTON,
ROBERT W. HUNT,
F. W. DEAN,
J. S. COON,
WM. B. POTTER,
} *Committee.*

RULES FOR CONDUCTING BOILER-TRIALS,

CODE OF 1897.

PRELIMINARIES TO A TRIAL.

I. *Determine at the outset* the specific object of the proposed trial, whether it be to ascertain the capacity of the

---

[*] The motion for the appointment of this Committee was made by Mr. Barrus in connection with the discussion of Mr. Dean's paper, No. DCL, on "The Efficiency of Boilers," etc. The President of the Society designated Mr. Kent, the chairman of the Committee of 1884, to call the first meeting of the new Committee. At that meeting, on motion of Mr. Kent, Dr. Emery was selected as chairman, and he conducted the preliminary correspondence. The report in the form originally printed was prepared by a sub-committee consisting of Messrs. Emery, Porter, Barrus, and Kent.

boiler, its efficiency as a steam-generator, its efficiency and its defects under usual working conditions, the economy of some particular kind of fuel, or the effect of changes of design, proportion, or operation; and prepare for the trial accordingly.

II. *Examine the boiler*, both outside and inside; ascertain the dimensions of grates, heating-surfaces, and all important parts; and make a full record, describing the same, and illustrating special features by sketches. The area of heating surface is to be computed from the outside diameter of all tubes, whether water-tubes or fire-tubes. This rule corresponds to the practice of many builders of different types of boilers, and is intended to make the practice of rating heating-surface uniform. All surfaces below the mean water-level which have water on one side and products of combustion on the other are to be considered as water-heating surface, and all surfaces above the mean water-level which have steam on one side and products of combustion on the other are to be considered as superheating surface.

III. *Notice the general condition* of the boiler and its equipment, and record such facts in relation thereto as bear upon the objects in view.

If the object of the trial is to ascertain the maximum economy or capacity of the boiler as a steam-generator, the boiler and all its appurtenances should be put in first-class condition. Clean the heating-surface inside and outside, remove clinkers from the grates and from the sides of the furnace. Remove all dust, soot, and ashes from the chambers, smoke-connections, and flues. Close air-leaks in the masonry and poorly fitted cleaning-doors. See that the damper will open wide and close tight. Test for air-leaks by firing a few shovels of smoky fuel and immediately closing the damper, observing the escape of smoke through the crevices.

IV. *Determine the character of the coal* to be used. For tests of the efficiency or capacity of the boiler the coal should, if possible, be of some kind which is commercially regarded as a standard. For New England and that portion of the country east of the Allegheny Mountains, good anthracite egg coal, containing not over 10 per cent of ash, and semi-bituminous Cumberland (Md.) and Pocahontas (Va.) coals are thus regarded. West of the Allegheny Mountains, Pocahontas (Va.) and New River (W. Va.) semi-bituminous, and Youghiogheny or Pittsburg bituminous coals are recognized as standards.* There is no special grade of coal mined in the Western States which is widely recognized as of superior quality or considered as a standard coal for boiler-testing. Big Muddy lump, an Illinois coal mined in Jackson County, Ill., is suggested as being of sufficiently high grade to answer the requirements in districts where it is more conveniently obtainable than the other coals mentioned above.

V. *Establish the correctness of all apparatus* used in the test for weighing and measuring. These are:

1. Scales for weighing coal, ashes, and water.
2. Tanks or water-meters for measuring water. Water-meters, as a rule, should only be used as a check on other measurements. For accurate work, the water should be weighed or measured in a tank.
3. Thermometers and pyrometers for taking temperatures of air, steam, feed-water, waste gases, etc.
4. Pressure-gauges, draft-gauges, etc.

The kind and location of the various pieces of testing apparatus must be left to the judgment of the person conducting the test, always keeping in mind the main object, i.e., to obtain authentic data.

VI. *See that the boiler and chimney are thoroughly heated* before the trial to their usual working temperature. If the

---

* These coals are selected because they are about the only coals which contain the essentials of excellence of quality, adaptability to various kinds of furnaces, grates, boilers, and methods of firing, and wide distribution and general accessibility in the markets.

boiler is new and of a form provided with a brick setting, it should be in regular use at least a week before the trial, so as to dry and heat the walls. If it has been laid off and become cold, it should be worked before the trial until the walls are well heated.

VII. *The boiler and connections* should be proved to be free from leaks before beginning a test, and all water connections, including blow and extra feed-pipes, should be disconnected, stopped with blank flanges, or bled through special openings beyond the valves, except the particular pipe through which water is to be fed to the boiler during the trial. During the test the blow-off and feed-pipes should remain exposed.

If an injector is used, it should receive steam directly through a felted pipe from the boiler being tested.*

See that the steam-main is so arranged that water of condensation cannot run back into the boiler.

VIII. *Starting and Stopping a Test.*—A test should last at least ten hours of continuous running. A longer test may be made when it is desired to ascertain the effect of widely varying conditions, or the performance of a boiler under the working conditions of a prolonged run. The conditions of the boiler and furnace in all respects should be, as nearly as possible, the same at the end as at the beginning of the test. The steam-pressure should be the same; the water-level the same; the fire upon the grates should be the same in quantity and condition; and the walls, flues, etc., should be of the same temperature. Two methods of obtaining the de-

---

* In feeding a boiler undergoing test with an injector taking steam from another boiler, or the main steam-pipe from several boilers, the evaporative results may be modified by a difference in the quality of the steam from such source compared with that supplied by the boiler being tested, and in some cases the connection to the injector may act as a drip for the main steam-pipe. If it is known that the steam from the main pipe is of the same quality as that furnished by the boiler undergoing the test, the steam may be taken from such main pipe.

sired equality of conditions of the fire may be used, viz.: those which were called in the Code of 1885 "the standard method" and "the alternate method," the latter being employed where it is inconvenient to make use of the standard method.

IX. *Standard Method.*—Steam being raised to the working pressure, remove rapidly all the fire from the grate, close the damper, clean the ash-pit, and as quickly as possible start a new fire with weighed wood and coal, noting the time and the water-level while the water is in a quiescent state, just before lighting the fire.

At the end of the test remove the whole fire, which has been burned low, clean the grates and ash-pit, and note the water-level when the water is in a quiescent state, and record the time of hauling the fire. The water-level should be as nearly as possible the same as at the beginning of the test. If it is not the same, a correction should be made by computation, and not by operating the pump after the test is completed.

X. *Alternate Method.* — The boiler being thoroughly heated by a preliminary run, the fires are to be burned low and well cleaned. Note the amount of coal left on the grate as nearly as it can be estimated; note the pressure of steam and the water-level, and note this time as the time of starting the test. Fresh coal which has been weighed should now be fired. The ash-pits should be thoroughly cleaned at once after starting. Before the end of the test the fires should be burned low, just as before the start, and the fires cleaned in such a manner as to leave the bed of coal of the same depth, and in the same condition, on the grates as at the start. The water-level and steam-pressures should previously be brought as nearly as possible to the same point as at the start, and the time of ending of the test should be noted just before fresh coal is fired. If the water-level is not the same as at the start, a correction should be made by computation, and not by operating the pump after the test is completed.

XI. *Uniformity of Conditions.*—In all standard trials the conditions should be maintained uniformly constant. Arrangements should be made to dispose of the steam so that the rate of evaporation may be kept the same from beginning to end. This may be accomplished in a single boiler by carrying the steam through a waste steam-pipe, the discharge from which can be regulated as desired. In a battery of boilers in which only one is tested the draught can be regulated on the remaining boilers, leaving the test-boiler to work under a constant rate of production.

Uniformity of conditions should prevail as to the pressure of steam, the height of water, the rate of evaporation, the thickness of fire, the times of firing and quantity of coal fired at one time, and as to the intervals between the times of cleaning the fires.

XII. *Keeping the Records.*—Take note of every event connected with the progress of the trial, however unimportant it may appear. Record the time of every occurrence and the time of taking every weight and every observation.

The coal should be weighed and delivered to the fireman in equal proportions, each sufficient for not more than one hour's run, and a fresh portion should not be delivered until the previous one has all been fired. The time required to consume each portion should be noted, the time being recorded at the instant of firing the last of each portion. It is desirable that at the same time the amount of water fed into the boiler should be accurately noted and recorded, including the height of the water in the boiler, and the average pressure of steam and temperature of feed during the time. By thus recording the amount of water evaporated by successive portions of coal, the test may be divided into several periods if desired, and the degree of uniformity of combustion, evaporation, and economy analyzed for each period. In addition to these records of the coal and the feed-water, half-hourly observations should be made of the temperature of the feed-water, of the flue gases, of the external air in the boiler-room, of the temperature of the furnace when a furnace-pyrometer is used, also of the pressure of steam, and of the readings of the instruments for determining the moisture in the steam. A log should be kept on properly prepared blanks containing columns for record of the various observations.

When the "standard method" of starting and stopping the test is used, the hourly rate of combustion and of evaporation and the horse-power may be computed from the records taken during the time when the fires are in active condition. This time is somewhat less than the actual time which elapses between the beginning and end of the run. This method of computation is necessary, owing to the loss of time due to kindling the fire at the beginning and burning it out at the end.

XIII. *Quality of Steam.*—The percentage of moisture in the steam should be determined by the use of either a throttling or a separating steam-calorimeter. The sampling-nozzle should be placed in the vertical steam-pipe rising from the boiler. It should be made of $\frac{1}{2}$-inch pipe, and should extend across the diameter of the steam-pipe to within half an inch of the opposite side, being closed at the end and perforated with not less than twenty $\frac{1}{8}$-inch holes equally distributed along and around its cylindrical surface, but none of these holes should be nearer than $\frac{1}{2}$ inch to the inner side of the steam-pipe. The calorimeter and the pipe leading to it should be well covered with felting. Whenever the indications of the throttling or separating calorimeter show that the percentage of moisture is irregular, or occasionally in excess of three per cent, the results should be checked by a steam-separator placed in the steam-pipe as close to the boiler as convenient, with a calorimeter in the steam-pipe just beyond the outlet from the separator. The drip from the separator should be caught and weighed, and the percentage of moisture computed therefrom added to that shown by the calorimeter.

Superheating should be determined by means of a thermometer placed in a mercury-well or oil-well inserted in the steam-pipe.

For calculations relating to quality of steam and corrections for quality of steam.

XIV. *Sampling the Coal and Determining its Moisture.*— As each barrow-load or fresh portion of coal is taken from the coal-pile, a representative shovelful is selected from it and placed in a barrel or box in a cool place and kept until the end of the trial. The samples are then mixed and broken into pieces not exceeding one inch in diameter, and reduced by the process of repeated quartering and crushing until a final sample weighing about five pounds is obtained, and the size of the larger pieces are such that they will pass through a sieve with $\frac{1}{4}$-inch meshes. From this sample two one-quart, air-tight glass preserving-jars, or other air-tight vessels which will prevent the escape of moisture from the sample, are to be promptly filled, and these samples are to be kept for subsequent determinations of moisture and of heating value, and for chemical analyses. During the process of quartering, when the sample has been reduced to about 100 pounds, a quarter to a half of it may be taken for an approximate determination of moisture. This may be made by placing it in a shallow iron pan, not over three inches deep, carefully weighing it, and setting the pan in the hottest place that can be found on the brickwork of the boiler setting or flues, keeping it there for at least twelve hours, and then weighing it. The determination of moisture thus made is believed to be approximately accurate for anthracite and semi-bituminous coals, and also for Pittsburg or Youghiogheny coal; but it cannot be relied upon for coals mined west of Pittsburg, or for other coals containing inherent moisture. For these latter coals it is important that a more accurate method be adopted. The method recommended by the Committee for all accurate tests, whatever the character of the coal, is described as follows:

Take one of the samples contained in the glass jars, crush the whole of it by running it through an ordinary coffee-mill adjusted so as to produce somewhat coarse grains (less than $\frac{1}{16}$ inch), thoroughly mix the crushed sample, select from it a portion of from 10 to 50 grams, weigh it in a balance which will easily show a variation as small as 1 part in 1000, and dry it in an air or sand bath at a temperature between 240 and 280 degrees Fahr. for one hour. Weigh it and record the loss, then heat and weigh it again repeatedly, at intervals of an hour or less, until the minimum weight has been reached and the weight begins to increase by oxidation of a portion of the coal. The difference between the original and the minimum weight is taken as the moisture. This moisture should preferably be made on duplicate samples, and the results should agree within 0.3 to 0.4 of one per cent, the mean of the two determinations being taken as the correct result.

If the coal contains an appreciable amount of surface moisture, another portion of the 100 pounds sample should be weighed and spread out in a thin layer on a clean sheet-iron plate, and exposed for a period of twenty-four hours to the atmosphere of the boiler-room, and by this means air-dried. After being weighed again, the percentage which the weight shrinks during this drying may be termed the percentage of surface moisture.

XV. *Treatment of Ashes and Refuse.*—The ashes and refuse are to be weighed in a dry state. For elaborate trials a sample of the same should be procured for analysis. When it is desired to know accurately the amount of coal consumed, as distinguished from combustible, all lumps of unconsumed coal one-half inch or more in diameter are to be picked from the refuse and deducted from the weight of coal fired.

XVI. *Calorific Tests and Analysis of Coal.*—The quality of the fuel should be determined either by heat test or by analysis, or by both.

The rational method of determining the total heat of combustion is to burn the sample of coal in an atmosphere of oxygen-gas, the coal to be sampled as directed in Article XIV of this Code.

The chemical analysis of the coal should be made only by an expert chemist. The total heat of combustion computed from the results of the ultimate analysis should be obtained by the use of Dulong's formula (with constants modified by recent determinations), viz.,

$$14600\ C + 62000 \left(H - \frac{O}{8} \right),$$

in which C, H, and O refer to the proportion of carbon, hydrogen, and oxygen respectively, and determined by the ultimate analysis.*

It is recommended that the analysis and the heat test be each made by two independent laboratories, and the mean of the two results, if there is any difference, be adopted as the correct figures.

It is desirable that a proximate analysis should also be made to determine the relative proportions of volatile matter and fixed carbon in the coal.

XVII. *Analysis of Flue-gases.*—The analysis of the flue-gases is an especially valuable method of determining the relative value of different methods of firing, or of different kinds of furnaces. In making these analyses great care should

---

* Favre and Silbermann give 14544 B. T. U. per pound carbon; Berthelot 14647 B. T. U. Favre and Silbermann give 62032 B. T. U. per pound hydrogen; Thomson, 61816 B. T. U.

be taken to procure average samples, since the composition is apt to vary at different points of the flue; and where complete determinations are desired, the analysis should be intrusted to an expert chemist. For approximate determinations the Orsat* or the Hempel† apparatus may be used by the engineer.

XVIII. *Smoke Observations.*—It is desirable to have a uniform system of determining and recording the quantity of smoke produced where bituminous coal is used. The system commonly employed is to express the degree of smokiness by means of percentages dependent upon the judgment of the observer. The Committee does not place much value upon a percentage method, because it depends so largely upon the personal element, but if this method is used, it is desirable that, so far as possible, a definition be given in explicit terms as to the basis and method employed in arriving at the percentage.

XIX. *Miscellaneous.*—In tests for purposes of scientific research, in which the determination of all the variables entering into the test is desired, certain observations should be made which are in general unnecessary for ordinary tests. These are the measurement of the air-supply, the determination of its contained moisture, the determination of the amount of heat lost by radiation, of the amount of infiltration of air through the setting, and (by condensation of all the steam made by the boiler) of the total heat imparted to the water.

As these determinations are not likely to be undertaken except by engineers of high scientific attainments, it is not deemed advisable to give directions for making them.

XX. *Calculations of Efficiency.*—Two methods of defining

---

* See R. S. Hale's paper on " Flue Gas Analysis," *Transactions A. S. M. E.*, vol. XVIII. p. 901.

† See Hempel on " Gas Analysis."

and calculating the efficiency of a boiler are recommended. They are:

1. Efficiency of the boiler $= \dfrac{\text{Heat absorbed per lb. combustible}}{\text{Heating value of 1 lb. combustible}}$

2. Efficiency of the boiler and grate
$= \dfrac{\text{Heat absorbed per lb. coal}}{\text{Heating value of 1 lb. coal}}$

The first of these is sometimes called the efficiency based on combustible, and the second the efficiency based on coal. The first is recommended as a standard of comparison for all tests, and this is the one which is understood to be referred to when the word "efficiency" alone is used without qualification. The second, however, should be included in a report of a test, together with the first, whenever the object of the test is to determine the efficiency of the boiler and furnace together with the grate (or mechanical stoker), or to compare different furnaces, grates, fuels, or methods of firing.

The heat absorbed per pound of combustible (or per pound coal) is to be calculated by multiplying the equivalent evaporation from and at 212 degrees per pound combustible (or coal) by 965.7.

In calculating the efficiency where the coal contains an appreciable amount of surface moisture, allowance is to be made for the heat lost in evaporating this moisture by adding to the heat absorbed by the boiler the heat of evaporation thus lost. The percentage of surface moisture used in this calculation is that which is found in the manner described in Article XIV of Code.

XXI. *The Heat-balance.*—An approximate "heat-balance," or statement of the distribution of the heating value of the coal among the several items of heat utilized and heat lost may be included in the report of a test when analyses of the fuel and of the chimney gases have been made. It should be reported in the following form:

HEAT BALANCE, OR DISTRIBUTION OF THE HEATING VALUE OF THE COMBUSTIBLE.

Total Heat Value of 1 lb. of Combustible.................. B. T. U.

	B. T. U.	Per Cent.
1. Heat absorbed by the boiler = evaporation from and at 212 degrees per pound of combustible $\times$ 965.7.		
2. Loss due to moisture in coal = per cent of moisture referred to combustible $\div$ 100 $\times$ [(212 $-$ t) + 966 + 0.48(T $-$ 212)](t = temperature of air in the boiler-room, T = that of the flue gases).		
3. Loss due to moisture formed by the burning of hydrogen = per cent of hydrogen to combustible $\div$ 100 $\times$ 9 $\times$ [(212 $-$ t) + 966 + 0.48(T $-$ 212)].		
4.* Loss due to heat carried away in the dry chimney gases = weight of gas per pound of combustible $\times$ 0.24 $\times$ (T $-$ t).		
5.† Loss due to incomplete combustion of carbon = $\dfrac{CO}{CO_2 + CO}$ $\times$ $\dfrac{\text{per cent C in combustible}}{100}$ $\times$ 10150.		
6. Loss due to unconsumed hydrogen and hydrocarbons, to heating the moisture in the air, to radiation, and unaccounted for.		
Totals....... ,..........		100.00

* The weight of gas per pound of carbon burned may be calculated from the gas analyses as follows:

Dry gas per pound carbon = $\dfrac{11\,CO_2 + 8\,O + 7(CO + N)}{3(CO_2 + CO)}$, in which CO_2, CO, O, and N are the percentages by volume of the several gases. As the sampling and analyses of the gases in the present state of the art are liable to considerable errors, the result of this calculation is usually only an approximate one. The heat-balance itself is also only approximate for this reason, as well as for the fact that it is not possible to determine accurately the percentage of unburned hydrogen or hydrocarbons in the flue gases.

The weight of dry gas per pound of combustible is found by multiplying the dry gas per pound of carbon by the percentage of carbon in the combustible, and dividing by 100.

† CO_2 and CO are respectively the percentage by volume of carbonic acid and carbonic oxide in the flue gases. The quantity 10150 = No. heat-units generated by burning to carbonic acid one pound of carbon contained in carbonic oxide.

XXII. *Report of the Trial.*—The data and results should be reported in the manner given in the following table, omitting lines where the tests have not been made as elaborately as provided for in such table. Additional lines may be added for data relating to the specific object of the test. The extra lines should be classified under the headings provided in the table, and numbered, as per preceding line, with sub letters, *a*, *b*, etc.

DATA AND RESULTS OF EVAPORATIVE TRIALS.

Made by................of.................boiler at...............to determine..
....
Principal conditions governing the trial......................................
....
....
Kind of fuel..
State of the weather..
   1. Date of trial...
   2. Duration of trial.. hours.

Dimensions and Proportions.

(A complete description of the boiler should be given on an annexed sheet.)

3. Grate surface........width.......length.......area...... sq. ft.
4. Water-heating surface..................................... "
5. Superheating surface....................................... "
6. Ratio of water heating surface to grate surface............
7. Ratio of minimum draft area to grate surface..............

Average Pressures.

8. Steam-pressure by gauge................. .... lbs.
9. Atmospheric pressure by barometer....................... in.
10. Force of draft between damper and boiler................. "
11. Force of draft in furnace................................. "
12. Force of draft in ash-pit................................. "

Average Temperatures.

13. Of external air... deg.
14. Of fire room... "
15. Of steam.. "
16. Of feed water entering heater............................ "
17. Of feed water entering economizer....................... "
18. Of feed water entering boiler............................ "
19. Of escaping gases from boiler........................... "
20. Of escaping gases from economizer................ ..... "

Fuel.

21. Size and condition..
22. Weight of wood used in lighting fire.................... lbs.
23. Weight of coal as fired *................................. "
24. Percentage of moisture in coal †....................... per cent.
25. Total weight of dry coal consumed (Art. XIV, Code)...... lbs.
26. Total ash and refuse...................................... "
27. Total combustible consumed............................ "
28. Percentage of ash and refuse in dry coal................ per cent.

Proximate Analysis of Coal.

	Of Coal.	Of Combustible.
29. Fixed carbon................................	per cent.	per cent.
30. Volatile matter.............................	"	"
31. Moisture....................................	"	—
32. Ash..	"	—
	100 per cent.	100 per cent.
33. Sulphur, separately determined.........	"	"

Ultimate Analysis of Dry Coal.
(Art. XVI, Code.)

34. Carbon (C).. per cent.
35. Hydrogen (H).. '
36. Oxygen (O).. "
37. Nitrogen (N).. "
38. Sulphur (S)... "
                                                            100 per cent.
39. Moisture in sample of coal as received................. "

Analysis of Ash and Refuse.

40. Carbon... per cent.
41. Earthy matter.. "

Fuel per Hour.

42. Dry coal consumed per hour............................. lbs.
43. Combustible consumed per hour........................ "
44. Dry coal per square foot of grate surface per hour...... "
45. Combustible per square foot of water heating surface per hour... "

---

* Including equivalent of wood used in lighting the fire, not including unburnt coal withdrawn from furnace at end of test. One pound of wood is taken to be equal to 0.4 pound of coal.

† This is the total moisture in the coal as found by drying it artificially, as described in Art. XIV of Code.

Calorific Value of Fuel.

46. Calorific value by oxygen calorimeter, per pound of dry coal.. B. T. U.
47. Calorific value by oxygen calorimeter, per pound of combustible... " " "
48. Calorific value by analysis, per lb. of dry coal*........... " " "
49. Calorific value by analysis, per pound of combustible..... " " "

Quality of Steam.

50. Percentage of moisture in steam......................... per cent.
51. Number of degrees of superheating...................... deg.
52. Quality of steam (dry steam = unity) ....
53. Factor of correction for quality of steam (page 119).......

Water.

54. Total weight of water fed to boiler....................... lbs.
55. Water actually evaporated, corrected for quality of steam "
56. Equivalent water evaporated into dry steam from and at degrees... "

Water per Hour.

57. Water evaporated per hour, corrected for quality of steam "
58. Equivalent evaporation per hour from and at 212 degrees. "
59. Equivalent evaporation per hour from and at 212 degrees per square foot of water-heating surface.............. "

Horse-power.

60. Horse-power developed. (34½ lbs. of water evaporated per hour into dry steam from and at 212 degrees, equals one horse-power)†..................................... H. P.
61. Builders' rated horse power.............................. "
62. Percentage of builders' rated horse-power developed...... per cent.

Economic Results.

63. Water apparently evaporated per lb. of coal under actual conditions. (Item 54 ÷ Item 23)...................... lbs.
64. Equivalent evaporation from and at 212 degrees per lb. of coal (including moisture)............................. "
65. Equivalent evaporation from and at 212 degrees per lb. of dry coal.. "
66. Equivalent evaporation from and at 212 degrees per lb. of combustible.. "

---

* See formula for calorific value under Article XVI of Code.

† Held to be the equivalent of 30 lbs. of water per hour evaporated from 100 degrees Fahr into dry steam at 70 lbs. gauge-pressure (See Introduction to Code.)

Efficiency.

(See Art. XX, Code.)

67. Efficiency of the boiler; heat absorbed by the boiler per lb. of combustible divided by the heat-value of one lb. of combustible.* .... per cent.
68. Efficiency of boiler, including the grate; heat absorbed by the boiler, per lb. of dry coal fired, divided by the heat value of one lb. of dry coal.† ....

Cost of Evaporation.

69. Cost of coal per ton of 2240 lbs. delivered in boiler-room... $
70. Cost of fuel for evaporating 1000 lbs. of water under observed conditions .... $
71. Cost of fuel used for evaporating 1,000 lbs. of water from and at 212 degrees .... $

Smoke Observations.

72. Percentage of smoke as observed ....
73. Weight of soot per hour obtained from smoke-meter ....
74. Volume of soot obtained from smoke-meter per hour ....

---

* In all cases where the word "combustible" is used, it means the coal without moisture and ash, but including all other constituents. It is the same as what is called in Europe "coal dry and free from ash."

† The heat value of the coal is to be determined either by an oxygen calorimeter or by calculation from ultimate analysis. When both methods are used the mean value is to be taken.

TABLE I.—HEAT OF COMBUSTION OF SUBSTANCES.

	Calories.	B. T. U.	
Crystallized carbon to CO_2	7859	14146	Berthelot
" " to CO	2405	4329	"
Amorphous carbon to CO_2	8137	14647	"
" " to CO	2489	4480	"
Graphite to CO_2	7901	14222	"
Petroleum coke to CO_2	8017	14503	Mahler
Gas coke to CO_2	8047	14485	F. & S.
Carbon vapor to CO_2	8722	15700	Calculated. Page 173.
Coal (pure and dry)	7800 to 9000	14040 to 16200	Various
Lignite (pure and dry)	6000 to 7000	10800 to 12600	"
Beech charcoal	7140	12852	Schwackhöfer
Soft charcoal	7071	12723	"
Cellulose	4200	7560	Berthelot
Soft resinous wood	5050	9090	Gottlieb
Hard wood	4750	8550	"
Peat	5940	10692	Bainbridge
Cane sugar	3961	7130	Berthelot
Asphalt	9532	17159	Slosson & Colburn
Pitch	8400	15120	Anon.
Naphthalin	9690	16842	Berthelot
Paraffin	11000	19800	Mahler
Tallow	9500	17100	Stohmann
Sulphur	2500	4500	Berthelot
Petroleum	9600 to 11000	17280 to 19800	Various
Schist-oil	9000 to 10000	16200 to 18000	"
Heavy coal gas oil	8900	16020	Ste-Claire Deville
Cotton oil	9500	17100	Anon.
Rape oil	9489	17080	Stohmann
Olive oil	9473	17051	"
Sperm oil	10000	18000	Gibson
Hydrogen	34500	62100	Berthelot
Carbonic oxide	2435	4383	"
Marsh gas	13343	24017	"
Olefiant gas	12182	21898	"
Acetylene	12142	21856	"
Carbon vapor (diamond)	11134	20041	"
Coal gas	4440 to 7370	7990 to 12266	Various
Petroleum gas	10800	19440	Anon.
Air producer gas	773 to 1370	1391 to 2466	Various
Water gas	2350 to 3032	4230 to 5458	"
Mixed gas	1015 to 1548	1827 to 2786	"

TABLE II.—THERMOMETER REDUCTION TABLES.

A. Centigrade to Fahrenheit.

C.	F.	C.	F.	C.	F.	C.	F.
1	1.8	10	18	100	180	1000	1800
2	3.6	20	36	200	360	2000	3600
3	5.4	30	54	300	540	3000	5400
4	7.2	40	72	400	720	4000	7200
5	9.0	50	90	500	900	5000	9000
6	10.8	60	108	600	1080	6000	10800
7	12.6	70	126	700	1260	7000	12600
8	14.4	80	144	800	1440	8000	14400
9	16.2	90	162	900	1620	9000	16200

B. Fahrenheit to Centigrade.

F.	C.	F.	C.	F.	C.	F.	C.
1	$\frac{5}{9}$	10	$5\frac{5}{9}$	100	$55\frac{5}{9}$	1000	$555\frac{5}{9}$
2	$1\frac{1}{9}$	20	$11\frac{1}{9}$	200	$111\frac{1}{9}$	2000	$1111\frac{1}{9}$
3	$1\frac{2}{3}$	30	$16\frac{2}{3}$	300	$166\frac{2}{3}$	3000	$1666\frac{2}{3}$
4	$2\frac{2}{9}$	40	$22\frac{2}{9}$	400	$222\frac{2}{9}$	4000	$2222\frac{2}{9}$
5	$2\frac{7}{9}$	50	$27\frac{7}{9}$	500	$277\frac{7}{9}$	5000	$2777\frac{7}{9}$
6	$3\frac{1}{3}$	60	$33\frac{1}{3}$	600	$333\frac{1}{3}$	6000	$3333\frac{1}{3}$
7	$3\frac{8}{9}$	70	$38\frac{8}{9}$	700	$388\frac{8}{9}$	7000	$3888\frac{8}{9}$
8	$4\frac{4}{9}$	80	$44\frac{4}{9}$	800	$444\frac{4}{9}$	8000	$4444\frac{4}{9}$
9	5	90	50	900	500	9000	5000

Having given Centigrade degrees, obtain from Table A the corresponding equivalents, and to their sum add 32°.

Example: Find Fahrenheit degrees corresponding to 416° C.

$$720 + 18 + 10.8 + 32 = 780.8.$$

Having given Fahrenheit degrees, subtract 32° *and find the value in Table B corresponding to the remainder.*

Example: Find Centigrade degrees corresponding to —16° F.

$$-16 - 32 = -48, \quad -48° F. = -(22\tfrac{2}{9} + 4\tfrac{4}{9}) = -26\tfrac{2}{3}.$$

TABLE III.—THEORETICAL FLAME TEMPERATURES.

	In Oxygen.		In Air.	
	Centigrade.	Fahrenheit.	Centigrade.	Fahrenheit.
C to CO.........................	4265°	7677°	1462°	2639°
C to CO_2......................	10000	18000	2718	4892
CO to CO_2	7010	12618	3000	5400
Hydrogen........................	6727	12108	2674	4813
Marsh gas, CH_4................	7971	14348	2245	4036
Olefiant gas, C_2H_4............	9659	17286	3000	5400
Acetylene, C_2H_2...............	11300	20340	3400	6120
Benzin, C_6H_6..................	9350	16830	2790	5022
Producer gas....................	2500	4500	1200	2160
Coal gas........................	5400	9720	2700	4860
Petroleum	7558	13604	2400	4320
Naphthalin	9444	17000	2730	4914
Wood	5800	10440	2280	4104
Lignite (dry)...................	3000	5400	1200	2160
Coal (bituminous)...............	3800	6840	1500	2700
Sulphur to H_2SO_4.............	2300	4140	1060	1908

TABLE IV.—WEIGHT AND VOLUME OF GASES.

Name.	Weight.		Volume.	
	Per Cubic Metre in Kilograms.	Per Cubic Foot in Pounds.	Per Kilogram in Cubic Metres.	Per Pound in Cubic Feet.
Air...................	1.29318	0.08073	0.773	12.385
Nitrogen..............	1.25616	0.07845	0.796	12.763
Oxygen...............	1.4298	0.08926	0.699	11.203
Hydrogen.............	0.08961	0.00559	11.160	178.83
Carbonic acid.........	1.9666	0.12344	0.508	8.147
Carbonic oxide........	1.2515	0.07817	0.800	12.800
Carbon vapor.........	1.0727	0.06696	0.932	14.930
Aqueous vapor........	0.8047	0.05022	1.242	19.912
Sulphurous acid.......	2.8605	0.1787	0.349	5.596
Ethylene, C_2H_4.......	1.2519	0.07814	0.799	12.797
Methane, CH_4	0.7155	0.04466	1.397	22.391
Acetylene, C_2H_2.....	1.1900	0.07428	0.840	13.456
Benzine, C_6H_6........	3.3333	0.208	0.303	4.808
Ethane, C_2H_6	1.3415	0.08565	0.746	11.950

TABLE V.—WEIGHT AND VOLUME OF OXYGEN AND AIR NECESSARY FOR COMBUSTION (Ser).

Combustibles.	Molecular Weights.			Weight per Kilogram of Combustible.				Composition by Volume.			Volume in Cubic Metres at 0° per Kilogram of Combustible.				
	Combustible.	Oxygen.	Products.	Combustion.				Combustible.	Oxygen.	Products.	Gaseous Combustible.	Combustible.			
				By Oxygen.		By Air.						By Oxygen.		By Air.	
				Oxygen.	Products.	Air.	Products.	Vol.	Vol.	Vol.		Oxygen.	Products.	Air.	Products.
				Kilo.	Kilo.	Kilo.	Kilo.								
Carbon.........	12	32	$CO_2 = 44$	2.667	3.667	11.594	12.594	1C	2O	$2CO_2$	0.9325	1.8650	1.8650	8.9669	8.9669
Carbon.........	12	16	$CO = 28$	1.333	2.333	5.797	6.797	1C	1O	$2CO$	0.9325	0.9325	1.8650	4.4834	5.4159
CO.............	28	16	$CO_2 = 44$	0.571	1.571	2.484	3.484	2CO	1O	$2CO_2$	0.7986	0.3993	0.7986	1.9188	2.3181
Hydrogen......	2	16	$H_2O = 18$	8.000	9.000	34.784	35.784	2H	1O	$2H_2O$	11.1700	5.5850	11.1700	26.8500	32.4350
Methane, CH_4...	16	$8O = 128$	$CO_2 = 44$ $2H_2O = 36$	4.000	5.000	17.392	18.392	$1C+4H$	4O	$2CO_2$ $4H_2O$	1.3990	2.7980	4.1970	13.4520	14.8510
Ethylene, C_2H_4..	28	$12O = 192$	$2CO_2 = 88$ $2H_2O = 36$	3.428	4.428	14.903	15.903	$1C+2H$	3O	$2CO_2$ $2H_2O$	0.7986	2.3958	3.1940	11.5190	12.3176

TABLE VI.—WEIGHT AND VOLUME OF OXYGEN AND AIR NECESSARY FOR COMBUSTION (Ser,' reduced.)

Combustibles.	Molecular Weights.			Weight per Pound of Combustibles.				Composition by Volume.				Volume in Cubic Feet at 32°, per Pound of Combustible.					
				Combustion.											Combustible.		
	Combustible.	Oxygen.	Products.	By Oxygen.		By Air.		Combustible.		Oxygen.		Product.		Gaseous Combustibles.	By Oxygen.		By Air.	
				Oxygen.	Products.	Air.	Products.	Vol.		Vol.		Vol.			Oxygen.	Product.	Air.	Product.
				Lbs.	Lbs.	Lbs.	Lbs.							Cubic Feet.	Cubic Feet.	Cubic Feet.	Cubic Feet.	Cubic Feet.
Carbon.......	12	32	CO$_2$ 44	2.667	3.667	11.594	12.594	1C		2		2CO$_2$		14.93	29.86	29.86	139.45	139.45
Carbon.......	12	16	CO 28	1.333	2.333	5.797	6.797	1C		1		2CO		14.93	14.93	29.86	69.72	84.65
Carb. ox., CO..	28	16	CO$_2$ 44	0.571	1.571	2.484	3.484	2CO		1		2CO$_2$		12.79	6.395	12.79	30.74	37.14
Hydrogen	2	16	H$_2$O 18	8.000	9.000	34.784	35.784	2H		1		2H$_2$O		178.94	89.47	178.94	430.14	519.61
Methane, CH$_4$.	16	8O=128	CO$_2$ 44 H$_2$O 36	4.000	5.000	17.392	18.392	1C,4H		4		2CO$_2$ 4H$_2$O		22.41	44.82	67.23	215.50	237.91
Ethylene,C$_2$H$_4$	28	12O=192	2CO$_2$ 88 2H$_2$O 36	3.426	4.428	14.903	15.903	1C,2H		3		2CO$_2$ 2H$_2$O		12.79	38.37	51.16	184.53	197.32

TABLE VII.—HEAT OF COMBUSTION OF GASES BY WEIGHT AND VOLUME.

At 0° and 760 mm. (32° F. and 29.922 in.). Berthelot.

Name of Gas.	Formula $H=1$ $C=12$	Composition by Volume.	Heat of Combustion.							
			Water Liquid.			Water Vapor.			
			Per Kilogram.	Per Cubic Metre.	Per Pound.	Per Cubic Foot.	Per Kilogram.	Per Cubic Metre.	Per Pound.	Per Cubic Foot.
Hydrogen.........	H	H_2	34,500	3,091	62,100	347	29,100	2,663	52,380	293
Marsh gas........	CH_4	1 vol. C + 4 vol. H	13,343	10,038	24,017	1,073	11,993	8,585	22,587	1,024
Acetylene........	C_2H_2	1 vol. C + 1 vol. H	12,142	14,460	21,856	1,624	11,727	13,961	21,109	1,569
Ethylene.........	C_2H_4	1 vol. C + 2 vol. H	12,182	15,250	21,898	1,712	11,411	14,261	20,509	1,503
Ethane...........	C_2H_6	1 vol. C + 3 vol. H	12,410	16,641	22,338	1,912	11,330	15,120	20,394	1,746
Benzene Vapor....	C_6H_6	1 vol. C + 1 vol. H	10,052	33,496	18,094	4,208	9,637	31,800	17,347	4,033
Carbonic Oxide...	CO	1 vol. C + 1 vol. O	2,435	3,043	4,383	341	......	....	....	....
Carb. Vap. (diamond)	C	C_4 (?)	11,734	11,935*	20,041	1,342	......	....	....	....
Carb. Vap. (charcoal)	C		11,328	12,143	20,390	1,366	......	....	....	....

* Carbon vapor (heat of vaporization not included) 8424 and 8722.

TABLE VIII.—SPECIFIC HEAT OF GASES, REFERRED TO WATER.

		Molecular Specific Heat in Grammes. C	Molecular Weight. M	Specific Heat by Weight. $\dfrac{C}{M}$	Weight per Cubic Metre at 0° and 0.760 m. P	Specific Heat by Volume per Cubic Metre. $\dfrac{C}{M}P$
		At 100°		Kilo.	Kilo.	Kilo.
Air.........				0.237	1.293	0.306
Nitrogen.....	N	100° = 6.83	$N_2 = 28$	0.244	1.256	0.306
Carbonic acid....	CO_2	100° = 9.54	$CO_2 = 44$	0.217	1.966	0.426
Carbonic oxide.....	CO	100° = 6.86	$CO = 28$	0.245	1.251	0.306
Olefiant gas.......	C_2H_4	100° = 11.72	$C_2H_4 = 28$	0.418	1.252	0.523
Marsh gas......	CH_4	100° = 9.49	$CH_4 = 16$	0.593	0.715	0.424
Hydrogen.....	H	100° = 6.82	$H_2 = 2$	3.410	0.0896	0.304
Oxygen......	O	100° = 6.95	$O_2 = 32$	0.217	1.430	0.310
Sulphurous acid.....	SO_2	100° = 9.86	$SO_2 = 48$	0.154	2.860	0.440
Aqueous vapor.....	H_2O	100° = 8.65	$H_2O = 18$	0.479	0.804	0.385

TABLE IX.—TABLE OF SPECIFIC HEAT OF GASEOUS PRODUCTS OF COMBUSTION REFERRED TO THE PROPORTION OF CARBONIC ACID.

Proportion of Carbonic Acid.	Specific Heat.	Proportion of Carbonic Acid.	Specific Heat.
5 per cent	0.312	11 per cent	0.319
6 " "	0.314	12 " "	0.320
7 " "	0.315	13 " "	0.321
8 " "	0.316	14 " "	0.322
9 " "	0.317	15 " "	0.323
10 " "	0.318		

TABLE X.—HEAT OF VAPORIZATION OF WATER AT 0° TO 230° C.

Temperature.		Heat of Vaporization.
Centigrade.	Fahrenheit.	
0	32	606.5
100	212	537.0
230	456	676.6

Latent heat of vaporization, 966 (Regnault).

TABLE XI.—SPECIFIC HEAT OF WATER (REGNAULT).

Temperature.	Specific Heat.	Temperature.	Specific Heat.
0°	1.0000	110°	1.0153
10	1.0005	120	1.0177
20	1.0012	130	1.0204
30	1.0020	140	1.0232
40	1.0030	150	1.0262
50	1.0042	160	1.0294
60	1.0056	170	1.0328
70	1.0072	180	1.0364
80	1.0098	190	1.0401
90	1.0109	200	1.0440
100	1.0130		

TABLE XII.—VOLUME OF OXYGEN TO FORM WATER WITH THE HYDROGEN OF COAL.

Per Cent of Hydrogen.	Oxygen in Litres per Kilogram of Coal.	Oxygen in Cubic Feet per Pound of Coal.
1	55.9	.896
2	112	1.792
3	168	2.699
4	223	3.585
5	279	4.481
6	335	5.397
7	391	6.283
8	446	7.170
9	502	8.096

TABLE XIII.—QUANTITY OF AIR REQUIRED FOR PERFECT COMBUSTION OF FUELS.

Fuel.	Composition.				Air per—	
	Carbon.	Hydrogen.	Oxygen.	Nitrogen.	Kilogram.	Pound.
					cu. metres	cu. feet
Coke	98.0	0.5			10.09	162.06
Coal, anthracite	95.4	2.2	1.8	0.5	9.01	144.60
bituminous	87.0	5.0	4.0		8.93	143.40
coking	85.0	5.0	6.0		8.68	139.41
cannel	84.0	6.0	8.0		8.79	141.07
smithy	77.0	5.0	15.0		7.67	123.15
Charcoal	90.0	2.0			8.53	133.90
Lignite	71.0	5.0	19.0		7.02	112.43
Peat, dry	58.0	6.0	30.0		5.75	92.36
Wood, dry	50.0	6.0	42.0	1.0	4.57	73.36
Petroleum	85.0	14.0	1.0		10.76	172.86
Natural gas	68.7	22.5	1.0	6.2	14.20	227.93
Coal gas	58.0	23.7	1.4	3.8	14.51	233.06
Water gas	34.0	5.9	43.0	3.4	3.16	50.70
Producer gas	1.0	5.0	21.0	69.0	.72	11.56

TABLE XIV.—RELATION BY WEIGHT AND VOLUME OF THE COMPONENTS OF AIR.

Air contains by volume:

Nitrogen	78.35
Oxygen	20.77
Aqueous vapor	0.84
Carbonic acid	0.04
	100.00

Deducting the carbonic acid and aqueous vapor, we have:

Nitrogen	By volume:	79.04	By weight:		76.83
Oxygen	" "	20.96	"	"	23.17
		100.00			100.00

Ratio of nitrogen to oxygen:

By volume, $\frac{N}{O} = 3.771$. By weight, $\frac{N}{O} = 3.32$.

Ratio of air to oxygen:

By volume, $\frac{Air}{O} = 4.771$. By weight, $\frac{Air}{O} = 4.315$.

Ratio of air to nitrogen:

By volume, $\frac{Air}{N} = 1.265$. By weight, $\frac{Air}{N} = 1.302$.

TABLE XV.—IGNITION POINT OF GASES (Mayer and Münch).[*]

Marsh gas, CH_4	667° C.
Ethane, C_2H_6	616
Propane, C_3H_8	547
Acetylene, C_2H_2	580
Propylene, C_3H_6	504

[*] Berichte der deutscher Gesellschaft XXVI, 2421.

FUEL TABLES.

These tables contain all the available information covering the data required which have been published to date. They contain analyses of the fuels, and the heat units as determined by the authors, whose names are given. In some cases it has been necessary to recalculate the results as published by the experimenters to conform with the standard adopted. This applies especially to the coals and solid fuels, the data for which are given based on pure dry coal, i.e., on the combustible present. If the actual test of the sample as given is desired, it will be easy to make the necessary deductions. Some of the cokes and some of the natural gases have been calculated, the calculated results being within the limits of experimental error in these cases.

COAL.

COLORADO.

Name or Location.	Carbon.			Hydrogen.	Oxygen, Nitrogen, Sulphur.	Oxygen.	Nitrogen.	Sulphur.	Water.	Ash.	Heat Units of Combustible.		Authority.
	Fixed.	Volatile	Total.								Calories	B.T.U.	
Palisade	54.5	36.1	90.6							9.4	8168	14700	Carpenter
Diamond, Jerome Park	54.1	40.9	95.0							5.04	7666	13800	"
Union	54.0	37.5	91.5							8.5	8500	15300	"
New Castle (lump)	52.2	36.7	88.9							11.1	7638	13750	"
" (run of mine)	46.1	39.8	85.9							14.1	7778	14050	"
" "	53.3	38.8	92.1							7.9	8611	15500	"

ILLINOIS.

Name or Location.	Carbon.			Hydrogen.	Oxygen, Nitrogen, Sulphur.	Oxygen.	Nitrogen.	Sulphur.	Water.	Ash.	Heat Units of Combustible.		Authority.
	Fixed.	Volatile	Total.								Calories	B.T.U.	
Auburn (screenings)	47.3	37.5	84.8						7.4	15.2	7333	13200	Carpenter
Barclay	46.2	35.7							6.14	10.7	7159	12886	McConney
Blackheart	46.27	29.94							17.65	17.65	7465	13427	Forsyth
Bloomington	45.2	36.4							4.1	14.7	7228	13010	McConney
Bryant	42.64	32.91							2.42	22.03	7309	13157	Forsyth
Canton	46.69	36.99							3.51	12.81	7267	13080	"
Centralia	45.5	34.0							8.3	8.0	7727	13548	McConney
Claire	43.05	32.94							3.16	20.85	7301	13142	Forsyth
Cuba	48.62	36.38							4.17	10.83	7431	13376	"
Danville	45.4	43.7							4.8	6.2	7990	14382	McConney
"	46.4	37.1							5.6	10.9	7446	13403	"
Dumferline	45.57	32.91							2.46	19.06	7412	13342	Forsyth
Du Quoin	53.7	32.0							6.8	7.4	7775	13995	McConney
Edwards	43.62	34.48							1.90	20.0	7070	12726	Forsyth
Elmwood	35.41	27.69							1.36	35.54	7765	13978	"
Farmington	45.89	33.88							3.41	16.82	7293	13126	"
Gillespie	51.41	36.26								12.33	6921	12400	Carpenter

ILLINOIS—Continued.

Name or Location.	Carbon.			Hydrogen.	Oxygen, Nitrogen, Sulphur.	Oxygen.	Nitrogen.	Sulphur.	Water.	Ash.	Heat Units of Combustible.		Authority.
	Fixed.	Volatile	Total.								Calories	B.T.U.	
Ladd (lump)........	48.2	36.7								15.1	8168	14700	Carpenter
"	44.12	33.53								13.76	7709	13876	Forsyth
La Salle...........	44.0	39.4							8.49	8.4	7774	13994	McConney
Lincoln............	44.5	35.0							8.2	12.1	7202	12963	"
Morris.............	49.74	32.12							8.5	11.04	7251	13052	Forsyth
Mt. Olive..........	44.5	36.4							7.1	11.1	7310	13158	McConney
Mt. Pulaski........	46.5	35.8							8.0	11.0	7258	13064	"
Niantic............	47.4	36.3							7.7	10.0	7366	13259	"
Odin...............	50.9	34.0							7.9	8.5	7637	13747	"
Pana...............	46.9	36.4							6.1	9.1	7418	13352	Forsyth
Peoria.............	49.21	36.13							7.2	9.5	7402	13323	McConney
Peru...............	47.2	37.2							3.22	11.44	7557	13603	Forsyth
Pottstown..........	45.52	35.52							9.0	6.6	7397	13314	McConney
Riverton...........	48.4	35.4							4.58	14.38	7432	13378	Forsyth
St. David..........	46.2	34.6							6.4	9.8	7444	13399	McConney
Sandoval...........	50.9	35.1							2.0	17.2	7640	13753	"
Staunton (lump)....	48.0	36.0							7.2	6.9	7611	13700	Carpenter
Streator...........	44.05	39.19							4.45	16.0	7608	13694	Forsyth
" (lump).....	48.2	39.4								12.31	8000	14400	Carpenter
" (nut)......	54.5	35.6								12.4	7888	14400	"
" (screenings)	42.2	31.4								9.9	7750	14000	"
" "	45.4	39.6								26.4	7944	14300	"
" "	43.8	38.4								15.0	7833	14100	"
Vicary.............	47.14	36.60							5.78	17.8	7652	13774	Forsyth
Wilmington (lump)...	44.9	36.8								10.48	7806	14050	Carpenter
" (screenings).	39.7	32.6								13.3	7333	13200	"
" (washed scr'n'gs)	50.2	34.5								27.7	7888	14200	"
										15.3			

FUEL TABLES.

ILLINOIS.

Name or Location.	Carbon. Fixed.	Carbon. Volatile	Carbon. Total.	Hydrogen.	Oxygen, Nitrogen, Sulphur.	Oxygen.	Nitrogen.	Sulphur.	Water.	Ash.	Heat Units of Combustible. Calories	Heat Units of Combustible. B.T.U.	Authority.
Big Muddy.........	53.9	28.3						1.0	7.4	10.5	7754	13757	Engineers' Club of St. Louis.
"	53.7	30.1						1.2	6.1	9.2	7563	13613	"
"	55.7	31.8						2.9	5.9	6.6	7480	13464	"
Colchester........	44.8	25.0							11.6	18.6	7840	14113	"
" (slack).....	38.2	25.4						1.2	5.3	31.1	7893	14207	"
Collinsville.......	31.6	43.9						5.3	9.2	13.3	7276	13097	"
Dumferline (slack)..	39.5	28.9							9.6	22.0	7649	13768	"
Du Quoin..........	49.9	30.3						0.9	11.3	8.5	7417	13351	"
Gillespie..........	45.3	30.6						1.5	12.6	11.5	6950	12510	"
Girard............	45.8	34.4						3.5	9.7	10.2	6900	12420	"
"	42.9	32.2						8.1	8.9	16.0	7610	13698	"
Heitz Bluff.......	48.2	37.8						3.3	9.0	5.0	6675	12015	"
Oakland...........	43.1	34.4						4.4	8.3	14.2	7450	13410	"
St. Bernard.......	48.4	34.4						1.4	14.4	6.4	7066	12720	"
St. Clair..........	39.7	30.9						9.6	7.8	21.8	7209	12976	"
"	41.8	33.1						6.9	10.3	14.9	7636	13746	"
St. John..........	44.9	34.2						4.3	11.2	9.7	7476	13457	"
"	45.8	28.4						2.1	9.8	16.1	7321	13180	"
Streator...........	43.5	24.5						1.8	13.6	15.4	7690	13840	"
"	48.8	35.3						2.4	12.1	3.9	7554	13580	"
Trenton...........	52.0	30.4						0.9	13.3	4.3	7139	12850	"
"	52.0	31.0						1.0	9.9	7.0	7520	13535	"
Vulcan (nut)......	45.1	30.9						1.3	7.4	16.6	6908	12435	"
"	50.0	27.9						0.7	10.3	12.8	7677	13817	"

NEW MEXICO.

Name or Location.	Carbon. Fixed.	Carbon. Volatile	Sulphur.	Water.	Ash.	Calories	B.T.U.
Location unknown....	50.2	35.5	0.6	2.4	11.9	7610	13700

COAL. 213

Name or Location.	Carbon. Fixed.	Carbon. Volatile.	Carbon. Total.	Hydrogen.	Oxygen, Nitrogen, Sulphur.	Oxygen.	Nitrogen.	Sulphur.	Water.	Ash.	Heat Units of Combustible. Calories	Heat Units of Combustible. B.T.U.	Authority.
INDIANA.													
Brazil................	50.30	34.49	70.50	4.76		16.29	1.36	1.39	8.98	6.28	8079	14542	Noyes, McTaggart and Craven
Caseyville............	44.49	30.67	57.37	5.01					1.51	23.69	7714	13885	Johnson
Lancaster.............	47.22	37.44	71.41	5.56		12.78	1.54	0.62	12.66	2.68	7917	14251	N. McT. & C.
New Pittsburg........	39.93	39.92	62.88	5.07		18.42	1.01	7.46	6.83	13.30	7732	13917	"
"	40.40	42.23	65.26	5.17		13.06	1.17	5.88	5.89	11.48	7763	13973	"
Shelburn..............	43.45	38.82	66.86	5.30		13.25	1.50	2.57	8.63	9.05	7935	14283	"
						15.69							
INDIAN TERRITORY.													
Atoka.................	57.32	35.42						3.73	6.66	6.60	6160	11088	Anonymous
Choctaw Nation........	66.85	23.31						1.18	1.59	8.25	7105	12789	
IOWA.													
Chisholm..............	39.58	40.42							9.18	10.82	7896	14213	Forsyth
Flagler's..............	37.69	40.16							9.48	12.31	7589	13661	"
Hiteman...............	25.37	35.27							4.99	34.37	8204	14767	"
Keb...................	44.75	37.49							9.81	7.95	7752	13955	"
KENTUCKY.													
Seatonville (lump).....	45.4	37.6								17.0	8050	14500	Carpenter
Vanderpool (lump).....	56.9	35.5								7.6	8000	14400	
MARYLAND.													
Atkinson and Kemptman.	83.2	16.8									6653	11975	Johnson (Kent)
Cumberland............	80.75	13.0	86.50	4.75		2.50			1.25	5.0	9067	16321	Isherwood
"										5.0		13529	Barrus
Easby's...............	83.6	16.4									6261	11270	Johnson (Kent)
Easby and Smith.......	82.7	17.3									6315	11367	"
Eureka................										5.1	7992	14386	Barrus
George's Creek........										6.1	8411	15140	"

FUEL TABLES.

MISSISSIPPI.

Name or Location.	Carbon.			Hydrogen.	Oxygen, Nitrogen, Sulphur.	Oxygen.	Nitrogen.	Sulphur.	Water.	Ash.	Heat Units of Combustible.		Authority.
	Fixed.	Volatile	Total.								Calories	B.T.U.	
Osage River............	51.16	41.83	81.85	6.20	1.98				1.67	5.34	8204	14768	Johnson

MISSOURI.

Brookfield............													
Hamilton.............	47.69	34.24							5.06	13.01	7508	13514	Forsyth
Lingo................	47.24	38.29							7.33	7.14	8157	14682	"
Mendota.............	46.24	37.48							9.03	7.25	7596	13673	"

NEBRASKA.

| Hastings............. | 60.88 | 27.82 | | | | | | | 0.21 | 11.09 | 8102 | 14583 | Forsyth |

OHIO.

Brier Hill............	59.1	36.4								4.5	7888	14200	Carpenter
Cambridge............	50.36	37.79	70.61	4.92		8.17	1.44	3.01	2.43	9.42	8041	14474	Lord and Haas
East Palestine.........	52.65	34.98	70.58	4.79		7.03	1.24	3.65	0.82	11.89	8113	14603	"
" "	51.32	37.45	73.23	4.97		7.35	1.47	1.75	1.65	9.58	8257	14863	"
Hocking Valley (lump)...	50.32	37.13	69.42	4.60		10.30	1.46	1.67	6.72	5.83	7870	14166	"
" " "	46.85	36.60						1.63	6.45	10.10	7913	14243	"
" " "	49.05	36.05	68.18	4.65		9.40	1.44	1.43	6.40	8.50	7767	13981	"
" " "	53.4	33.8								8.8	7833	14100	Carpenter
" " (run of mine)	49.54	34.14	66.50	4.42		10.66	1.43	1.67	6.65	9.67	7762	13972	Lord and Haas
" " "	47.95	35.18						1.50	6.34	10.53	7797	13314	"
Jackson..............	54.57	34.31							4.10	7.02	7753	13956	McConney
Palestine.............	50.70	36.70	71.29	4.76		7.37	1.34	2.64	2.15	10.45	8339	15010	Lord and Haas
"	52.65	36.60	73.64	4.79		7.29	1.24	2.34	2.45	8.25	8224	14803	"
Salineville (Mahoning)..	50.95	35.00	71.13	4.60		6.78	1.23	1.86	3.15	10.90	8182	14578	"
" (Freeport)...	52.80	36.30	72.62	4.82		7.43	1.23	3.00	2.80	9.10	8160	14688	"
Yellow Creek..........	50.88	38.72	73.15	4.84		6.32	1.40	3.89	1.23	9.17	8330	14994	"
Waterford............	53.34	37.29	74.39	4.98		6.42	1.40	3.44	1.55	7.82	8230	14814	"

PENNSYLVANIA (ANTHRACITE).

Name or Location.	Carbon.			Hydrogen.	Oxygen, Nitrogen, Sulphur.	Oxygen.	Nitrogen.	Sulphur.	Water.	Ash.	Heat Units of Combustible.		Authority.
	Fixed.	Volatile.	Total.								Calories	B.T.U.	
Avondale.............	87.78	6.00	93.89		Sp.gr. 1.44					6.91	7789	14200	Carpenter
Blackheath...........				3.55		2.56					8677	15619	Isherwood
Buck Mountain (slate out)	92.41	2.17			1.56					5.42	8333	15000	Carpenter
Cayuga, Scranton.....	84.38	6.5			1.49					9.12	7666	13800	"
Continental, Scranton..	84.19	5.78			1.61					10.03	7694	13850	"
Cross Creek..........										10.5	7151	12872	Barrus
Forty Foot, Scranton...	84.92	5.07			1.41					10.01	8056	14500	Carpenter
Honey Brook.........										12.00	7407	13332	Barrus
Jermyn (stove).......	82.9	6.08			1.425					11.02	7694	13850	Carpenter
Lackawanna..........	84.0	5.0								11.0	7724	13900	"
LykensValley (buckwheat)	76.94	6.21			1.30					17.5	7477	13459	Barrus
" "	80.2	6.8								15.5	7833	14100	Carpenter
" "	81.0	5.0								13.0	7600	13680	"
Mammoth, Drifton,										14.0	7583	13650	"
Buckwheat, slate out...	90.59	2.44			1.55					6.97	8194	14750	"
Manville Shaft, Scranton..	86.5	6.12			1.42					7.38	7833	14100	"
Mount Pleasant, Scranton.	81.59	7.63			1.42					10.78	7889	14200	"
" (pea)	76.28	7.49			1.41					10.01	7806	14050	"
Oxford...............	91.27	6.49		1.73	1.45	0.78	.001		0.84	2.24	7633	13740	"
Treverton............	85.66	6.67	90.66							6.83	8442	15195	Isherwood
Woodward, Scranton	81.87	4.06			1.42					14.07	8083	14551	Carpenter

216 FUEL TABLES.

PENNSYLVANIA (BITUMINOUS).

Name or Location.	Carbon.			Hydrogen	Oxygen, Nitrogen, Sulphur.	Oxygen.	Nitrogen.	Sulphur.	Water.	Ash.	Heat Units of Combustible.		Authority.
	Fixed.	Volatile	Total.								Calories	B.T.U.	
Antrim.............	70.16	18.54								11.30	8556	15400	McConney
Beaver Creek.......	55.42	34.33	74.60	4.89		6.90	1.40	1.96	1.50	8.75	8201	14762	Lord and Haas
Bernmont...........	59.96	32.00								8.04	8111	14600	McConney
Big Muddy	58.2	33.2								8.6	7981	14700	Carpenter
"	56.7	36.0								7.3	8000	14400	"
"	60.7	34.05								4.33	8444	15200	"
Carnegie...........	56.20	36.42	77.20	5.10		7.22	1.68	1.42	1.45.	5.93	8304	14947	Lord and Haas
"	55.06	37.79	76.57	5.01		7.85	1.64	1.76	1.07	6.08	8352	15034	"
"	52.00	37.67	73.50	5.01		7.12	1.44	2.54	1.08	9.25	8348	14846	"
Clinton............	53.80	35.60	73.57	4.86		7.87	1.24	1.86	2.55	8.05	8166	14699	"
Creedmoor.........	51.14	38.91	74.45	5.15		7.05	1.60	1.80	1.09	8.86	8324	14983	"
Eureka	70.39	23.79								5.82	8222	14800	McConney
Hoytdale No. 1*...	53.50	35.10	72.78	4.63		8.17	1.34	1.68	2.70	8.70	8177	14718	Lord and Haas
" No. 2...	57.65	36.40	77.83	5.04		7.96	1.65	1.57	1.60	4.35	8278	14900	"
Leisenring.........	64.49	29.26								6.25	9056	16300	McConney
New Galilee........	52.30	36.70	73.57	4.94		6.90	1.35	2.24	2.30	8.70	8165	14697	Lord and Haas
North Mansfield ...	52.65	36.20	73.91	4.92		7.02	1.23	1.77	2.10	9.05	8277	14898	"
Ormsby............	53.72	39.03	67.57	2.47		2.67	1.04		1.25	6.00	8333	14988	Isherwood
Pittsburgh.........	54.6	35.5								9.9	7889	14200	Carpenter
Reynoldsville......	69.96	24.67								5.37	8942	16100	McConney
Turtle Creek.......	56.59	34.38	76.56	5.10		6.04	1.67	1.60	1.08	7.95	8378	15080	Lord and Haas
"	53.00	36.20	74.48	4.86		6.83	1.37	1.66	1.75	9.05	8289	14919	"
Wampum No. 1*....	50.85	37.50	72.82	4.93		6.02	1.33	3.25	2.85	8.80	8267	14880	"
" No. 2	55.77	38.53	77.93	5.09		7.28	1.65	2.35	3.75	4.95	8256	14861	"
Near Wampum	55.85	36.80	76.81	5.14		7.90	1.62	1.18	0.70	6.65	8244	14839	"
Youghiogheny	54.7	32.6								5.9	7129	13752	Barrus
"	54.7	32.6								12.7	8333	15000	Carpenter
"	60.82	33.13							1.40	4.65	8906	16031	McConney

* Nos. 1 are from same mines as Nos. 2, but some years older.

COAL.

Name or Location.	Carbon. Fixed.	Carbon. Volatile	Carbon. Total.	Hydrogen.	Oxygen, Nitrogen, Sulphur.	Oxygen.	Nitrogen.	Sulphur.	Water.	Ash.	Heat Units of Combustible. Calories	Heat Units of Combustible. B.T.U.	Authority.
TENNESSEE.													
Glen Mary, Scott Co.....	61.63	31.47						0.94	2.15	4.75	7315	13167	Anonymous
TEXAS.													
Fort Worth............	49.27	34.72						1.56	4.60	11.41	6335	11403	Anonymous
Elser, Bowie Co.........											6244	11239	Slosson&Colburn
Webb, Bowie Co.........											6217	11190	"
UTAH.													
Castle Gate...........	51.3	41.8	93.1							6.9	7877	14180	Carpenter
WEST VIRGINIA.													
Barr's Deep Run.......	77.5	22.5									5805	10449	Johnson
Chesterfield Co.......	61.3	38.7									5665	10197	"
Clearfield............											7246	13043	Barrus
Clover Hill............	56.83	31.70	83.39	4.96		11.65			1.34	7.6	7925	14265	Johnson
Creek Co..............	63.8	36.2								10.13	5273	9491	"
Crouch and Snead.....	77.1	28.9									5594	10069	"
Eclipse...............										2.7	8060	14505	Barrus
Elenora...............										7.5–9.1	7667— 7918	13800— 14253	"
Elk Garden...........	55.9	37.5								7.8	7942	14295	"
Little Pittsburg........	56.4	33.49	93.62	5.74		0.64			0.67	6.6	7613	13704	Carpenter
Midlothian...........	64.3	35.7								9.44	9493	17087	Johnson
" (screenings)...											5708	10274	"
New River............										0.6–5.7	7077— 8112	14359— 14601	Barrus
Pocahontas (run of mine)..	73.65	18.30	83.75	4.13		2.65	0.85	0.57	0.80	7.25	8768	15682	Lord and Haas
"	73.60	17.05						0.60	0.75	8.60	8739	15730	"

WEST VIRGINIA—*Continued.*

Name or Location.	Carbon.			Hydrogen.	Oxygen, Nitrogen, Sulphur.	Oxygen.	Nitrogen.	Sulphur.	Water.	Ash.	Heat Units of Combustible.		Authority.
	Fixed.	Volatile	Total.								Calories	B.T.U.	
Pocahontas	75.12	18.62	85.46	4.18		2.66	0.85	0.57	0.63	5.63	8732	15718	Lord and Haas
"	74.48	17.92						0.63	0.61	6.99	8745	15728	"
"	75.75	18.60						0.62	0.65	4.80	8777	15798	"
" (average)	74.52	18.10	85.40	4.20		3.18	0.85	0.60	0.73	6.65	8751	15739	"
Sonman										6.20	8059	14509	Barrus
"										4.00	7986	14375	"
										6.9–8.3	7891—8037	14204—14467	
Thacker (nut 1)	55.45	36.07						1.81	1.18	7.30	8425	15165	Lord and Haas
" (nut 2)	56.25	36.35	78.40	5.04		6.36	1.40	1.40	1.35	6.05	8496	15293	"
" (run of mine 1)	57.10	35.00						1.16	1.40	6.50	8434	15181	"
" " 2)	56.15	34.75						1.18	1.60	7.50	8513	15323	"
" (average)	56.24	35.54	78.90	4.98		5.64	1.42	1.39	1.38	6.84	8467	15340	"

WYOMING.

Name or Location.	Fixed.	Volatile	Total.	Hydrogen.		Oxygen.	Nitrogen.	Sulphur.	Water.	Ash.	Calories	B.T.U.	Authority.
Almy, No. 7, upper vein	48.75	34.88	83.66*					0.65	7.37	9.00	7405	13330	Slosson&Colburn
" lower vein	51.75	33.55	85.30						8.82	5.90	7124	12824	"
Antelope, Cambria	44.25	39.38	83.63					3.79	6.72	9.65	7905	14230	"
Becker, Sheridan	38.75	35.25	74.00					1.04	14.10	11.90	7351	13232	"
Black Buttes No. 1	51.98	30.07	82.05					0.61	14.45	3.50	7342	13215	"
Bohack	42.60	39.25	81.85					0.80	13.65	4.50	6680	12022	"
Brown (1894)	45.00	36.85	81.85					1.13	11.25	6.90	7098	12777	"
" (1893)	47.30	34.65	81.95					1.25	11.85	6.20	7619	13715	"
Burgess, Sheridan	44.70	37.55	82.25					0.71	13.05	4.70	7063	12714	"
Caspar No. 1	53.55	32.10	85.65					0.40	11.30	3.20	7561	13610	"
Carbon No. 2	48.30	35.43	83.73						7.42	8.85	7823	14080	"

* Total fuel.

COAL. 219

WYOMING—*Continued.*

Name or Location.	Carbon.			Hydrogen.	Oxygen, Nitrogen, Sulphur.	Oxygen.	Nitrogen.	Sulphur.	Water.	Ash.	Heat Units of Combustible.		Authority.
	Fixed.	Volatile	Total.								Calories	B.T.U.	
Chase	44.75	34.50	75.25*					1.03	14.50	6.25	7309	13156	Slosson&Colburn
J. Curtis, Ham's Fork	57.75	37.90	95.65					1.00	1.50	2.85	7914	14246	" "
Deer Creek	47.75	33.03	80.73					0.61	13.82	5.40	6812	12262	" "
Diamond	44.30	33.35	77.65					0.42	14.50	7.85	6477	11658	" "
Earl and Gillis	48.00	34.25	82.25					0.71	13.25	4.50	6953	12515	" "
W. Goodell, Ham's Fork	54.00	38.00	92.00						2.95	4.05	8138	14648	" "
Grinnell	44.75	33.18	77.93					2.87	14.42	7.65	7074	12734	" "
Harker	43.90	33.52	77.40					1.03	7.88	14.70	7433	13380	" "
Holland, Buffalo	45.30	35.05	80.35						13.55	6.10	6701	12060	" "
Inez	42.50	36.05	79.15					0.66	14.65	6.80	6952	12513	" "
Jumbo	43.65	40.13	83.78					4.57	5.72	10.50	7873	14170	" "
A. Kendall, Ham's Fork	55.40	37.25	92.65					0.60	3.75	3.60	7983	14370	" "
Kindt No. 1	55.15	35.68	90.83					0.77	4.87	4.30	7896	14213	" "
Kindt No. 2	55.65	35.80	91.45					0.70	5.40	3.15	7801	14042	" "
Lander Fuel Co	47.60	36.60	84.20					0.50	11.40	4.40	7143	12857	" "
L. A. Mason	48.46	37.02	85.48					0.35	10.50	4.42	7055	12698	" "
McCoid	56.15	34.75	90.90					0.85	5.05	4.05	7652	13837	" "
Meyer, Carbon	53.00	33.10	83.10					0.65	11.15	2.75	7360	13248	" "
Monker and Mather	44.20	34.30	78.50					0.34	14.70	6.80	6452	11614	" "
New Dillon	54.25	33.25	87.50					0.50	7.25	4.25	7872	14170	" "
Red Canon No. 5	48.50	36.08	84.58					0.44	7.42	8.00	7555	13600	" "
Rock Springs (av. of 5)	55.60	33.58	89.18					0.83	7.17	3.65	7254	13057	" "
Sweetwater	55.70	36.95	92.65					0.86	5.55	1.80	7971	14348	" "
Van Dyke (lower vein)	56.50	34.50	91.00					0.74	6.25	2.75	7884	14190	" "
" (upper vein)	56.85	35.73	92.58					0.63	5.67	1.75	7518	13532	" "
E. L. Young, No. 1	50.85	31.51	82.36					0.56	14.66	1.98	7600	13680	" "

* Total fuel.

FUEL TABLES.

NOVA SCOTIA.

Name or Location.	Carbon. Fixed.	Carbon. Volatile.	Carbon. Total.	Hydrogen.	Oxygen, Nitrogen, Sulphur.	Oxygen.	Nitrogen.	Sulphur.	Water.	Ash.	Heat Units of Combustible. Calories	Heat Units of Combustible. B.T.U.	Authority.
Cooperstown	65.16	30.75								4.09	8944	16100	Carpenter
Cape Breton										7.8	6947	12505	Barrus
Nova Scotia	63.61	32.38								4.11	8889	16000	Carpenter

ONTARIO.

Name or Location.	Carbon. Fixed.	Carbon. Volatile.	Carbon. Total.	Hydrogen.	Oxygen, Nitrogen, Sulphur.	Oxygen.	Nitrogen.	Sulphur.	Water.	Ash.	Calories	B.T.U.	Authority.
Sudbury (Anthraxolite)			94.92			1.69	1.04	0.31	4.00	1.52	8141	14654	W. H. Ellis and W. Lawson

CHILI.

Name or Location.	Fixed.	Volatile.	Total.	Hydrogen.	O,N,S	Oxygen.	Nitrogen.	Sulphur.	Water.	Ash.	Calories	B.T.U.	Authority.
Cerro Verde, South Basin	47.40	41.35				0.60			11.25	4.75	7520	13536	Bobinski
Schwager	57.85	40.40				0.70			1.75	14.00	7891	14204	"
Coronel	61.40	39.25				0.85			6.25	7.00	6641	14954	"
Lota	62.30	40.00				0.93			7.00	10.40	7390	13302	"
Arauco	60.71	39.00				2.24			4.50	4.58	6532	11758	"
Lebu	59.57	42.00				0.83			7.50	8.67	7275	13095	"
Valdivia	56.14	45.00				0.58			8.25	9.15	7022	12640	"
Martha, Magellan	47.25	37.35				0.48			14.50	15.40	6983	12369	"
Madeleine	40.10	44.20				0.20			15.70	9.15	7971	14348	"
Punta Areñas	40.10	42.15				0.47			17.75	9.50	7917	14250	"

COAL.   221

FRANCE.

Name or Location.	Carbon.			Hydrogen.	Oxygen.	Oxygen, Nitrogen, Sulphur.	Nitrogen.	Water.	Ash.	Heat Units of Combustible.		Authority.
	Fixed.	Volatile	Total.							Calories	B.T.U.	
North Basin.												
Aniche................	84.8	4.6	89.4	4.0		6.6	0.6			8500	15300	S.-K.
Anzin, Fosse Lambrecht..	86.2	6.0	92.2	4.0		3.8	0.9			8600	15480	"
" " Lebret.........	77.2	7.3	84.5	4.2		11.3	0.7			9200	16560	"
" " St. Louis.....	82.2	1.8	84.0	3.7		11.6				8450	15210	"
" " St. Marc......			91.15	4.26		4.44	1.22			8656	15581	Mahler
Denam.................	70.4	13.6	84.0	4.4		11.6				9000	16200	S.-K.
Charleroi Basin.												
Bascoup...............			92.19	4.24		3.57				8820	15876	Bunte
".....................	84.42	7.66	92.08	6.04		1.88	0.84			8864	15955	S.-K.
".....................	82.79	2.11	84.90	4.58		10.52	0.69			8600	15480	"
Gilly-les-Charleroi....	86.71	3.71	90.46	3.76		5.78	0.65			8430	15174	"
Monceau-Fontaine......	82.71	0.57	83.28	3.98		12.74	0.89			8550	15390	"
Sart-les-Moulins......	84.13	3.01	87.14	6.31		6.55	0.84			8500	15300	"
".....................	85.74	7.97	93.71	4.10		2.19	0.73			8450	15210	"
										8430	15174	
Pas-de-Calais.												
Béthune...............	69.39	15.56	84.95	6.35		8.70	0.90			8360	15048	S.-K.
Courcelles-les-Lens	76.27	9.93	86.20	3.93		9.87	0.93			8650	15510	"
Courrières............	76.32	14.57	90.89	4.17		5.04	0.94			8800	15840	"
Dourges...............	78.60	13.15	91.75	3.13		5.12	0.91			8720	15696	"
".....................	78.29	3.98	82.27	4.98		12.75	0.89			8630	15534	"
".....................	75.74	6.01	81.75	5.41		12.84	0.79			8560	15408	"
Douvrin...............	86.48	1.07	87.55	3.77		8.68	0.60			8340	15012	S.-K. and M.-D.

FRANCE—*Continued.*

Name or Location.	Carbon.			Hydrogen.	Oxygen.	Oxygen, Nitrogen, Sulphur.	Nitrogen.	Water.	Ash.	Heat Units of Combustible.		Authority.
	Fixed.	Volatile	Total.							Calories	B.T.U.	
Pas-de-Calais.												
Fosse Douvrin, Lens	87.34	3.84	91.18	3.96		4.86	0.41			8640	15552	S.-K.
" " (slack)	72.25	14.30	86.55	3.83		9.62	0.79			8560	15408	"
Gas coal, Lens			87.26	5.43		7.30	1.04			8744	15739	Mahler
Meurchin	86.63	4.02	90.65	3.76		5.59	0.69			8440	15192	S.-K. and M.-D.
Commentry.												
Gas coal			85.66	5.60	8.20	8.73		3.45	7.05	8408	15134	Mahler
"			75.18	5.18						8305	14949	"
Creuzot (fat)	80.42	8.06	88.48	4.41		7.11	1.05			9400	16920	S.-K. and M.-D.
" (medium)	79.49	10.58	90.07	4.10		5.83	0.94			9200	16560	"
" (lean)	84.12	6.67	90.79	4.24		4.97				9040	16272	"
" (anthracite)	88.36	4.00	92.36	3.66		3.98				9200	16560	"
Blanzy Montceau	48.75	29.83	78.58	5.23		16.19				8100	15580	"
" anthracitic	74.54	12.48	87.02	4.71		8.26				8900	16020	"
St. Mary's Wells			34.26	5.27		10.40	1.20			8350	15030	Mahler
Treuil, St. Etienne (fat coal)			69.23	5.02		5.74	0.88			8856	15941	"
Ronchamp	72.57	17.39	89.96	5.09		4.95				8900	16020	S.-K. and M.-D.
"	74.74	12.75	87.49	5.15		7.41				8700	15660	"
"	71.44	15.99	87.43	4.56		8.01				8850	15930	"
"	71.58	16.80	88.38	4.42	3.49	7.20	1.30			8880	15984	"
"	78.81	10.28	89.09	5.09		5.82				8900	16020	S.-K.
"										8800	16200	"
" (average of 5)	73.83	14.62	88.45	4.86		6.68	1.25			8786	15816	S.-K. and M.-D.

GREAT BRITAIN.

Name or Location.	Carbon.			Hydrogen.	Oxygen.	Nitrogen.	Sulphur.	Water.	Ash.	Heat Units of Combustible.		Authority.
	Fixed.	Volatile	Total.							Calories	B.T.U.	
Bickershaw	63.87	29.90	78.93	4.90	7.24	1.56	1.07	4.36	1.96	7778	14000	Thwaite
Bwlf	82.01	9.07	91.08	3.83	12.45	5.09	1.45		3.31	8580	15274	S.-K. and M.-D.
Crombourke	57.87	31.67	69.77	4.82	12.45	1.33	1.01	7.15	2.65	8327	14989	Thwaite
Derbyshire (average of 8)	56.30	41.00	79.68	4.94	10.28	1.41	1.44		4.88	8120	14616	Anon.
Lancashire (average of 26)	53.12	42.00	77.90	5.32	9.53	1.30	1.44		3.77	8113	14602	"
Newcastle (average of 18)	57.23	39.00	82.12	5.31	5.69	1.35	1.24		4.91	8446	15203	"
Wales (average of 37)	68.09	27.00	83.78	4.79	4.15	0.98	1.43	1.62	3.22	8402	15123	"
Nixon Navigation	84.05	11.72	88.03	4.11	1.98	0.96	0.79	1.62		8340	15012	Thwaite
"	87.44	2.83	90 27	4.39	5.14	4.16	0.49	0.69		8864	15955	S.-K.
"			10.32									
Powel	87.72	5.14	92.49	4.04	3.47	1.41	1.02	6.70	4.55	8595	15471	Mahler
Pemberton	87.35	32.29	72.43	5.16	8.84	1.43	0.63	3.90	1.91	8750	15750	S.-K. and M.-D.
Pendleton	56.46	28.53	79.76	4.89	7.52	1.22	1.62	4.72	14.90	8161	14690	Thwaite
Thackerley Company	65.66	30.10	68.14	4.78	4.86	1.53	0.65	6.07	4.09	8213	14783	"
Tyldesley Company	50.28	32.09	74.46	5.10	8.25	1.59	0.75	2.78	6.55	8024	14433	"
Upper Drumgray Seam	57.76	25.63	75.48	4.98	7.86	1.44	1.24	4.64	2.75	7868	14162	"
Wigan	65.04	28.31	76.49	4.96	8.46					8145	14661	"
	64.10									8155	14679	"

AUSTRIA-HUNGARY.

Name or Location.	Fixed.	Volatile	Total.	Hydrogen.	Oxygen.	Nitrogen.	Sulphur.	Water.	Ash.	Calories	B.T.U.	Authority.
Dombrau, Austria			81.27	4.60	13.51	0.07	0.55			7921	14258	Schwackhöfer
Erzherzog Albrecht, Austria			82.75	4.69	11.40	0.37	0.79			8406	15131	"
Karwinerer-Larisch, Austria			83.00	4.70	11.43	0.34	0.50			8309	14956	"
Rossitz, Austria			82.47	4.68	12.10	0.18	0.57			7781	14006	"
Pilsen Priesen Kamotauer, Aus.			72.25	5.30	21.62	0.31	0.52			6978	12560	"
Witozek, Austria			82.50	4.82	12.03	0.23	0.42			8344	15054	"
Fünfkirchen, Hungary(av. of 4)			78.83	3 85	5.10	1.70		1.10		7373	13271	B. and H. Zeitung

AUSTRIA-HUNGARY—Continued.

Name or Location.	Carbon.			Hydrogen.	Oxygen.	Nitrogen.	Sulphur.	Water.	Ash.	Heat Units of Combustible.		Authority.
	Fixed.	Volatile	Total.							Calories	B.T.U.	
Szabolcs, Hungary (av. of 4)			79.98	4.10	4.75	1.42		1.33		7517	13530	B. and H. Zeitung
Vasos, Hungary (average of 3)			76.14	4.09	6.54	1.29		1.45		7162	12892	"
Carpano, Istria			66.94	4.08	17.54	1.28	8.72	2.01		6785	12213	Weithofer
Senge, Servia	46.88	36.15	58.12	3.78	20.73			13.32	14.15	7285	13113	"
" "	47.00	36.72	57.31	3.54	21.17			13.63	4.05	7054	12697	"
" "			59.85	4.44	19.41			12.63	2.65	7425	13365	"
" "			60.85	4.02	21.47			12.43	3.67	6903	12425	"
"Plattel-Kobele," Bohemia	51.70	34.64	79.38	7.36	13.26		1.43		1.23	8215	14787	Bunte

GERMANY.

Name or Location.	Carbon.			Hydrogen.	Oxygen.	Nitrogen.	Sulphur.	Water.	Ash.	Heat Units of Combustible.		Authority.
	Fixed.	Volatile	Total.							Calories	B.T.U.	
Hausham, Bavaria			57.61	4.43	18.84		1.99	7.75	11.37	7089	12760	Stuchlik
Penzberg, Bavaria			56.70	4.34	19.78		0.68	11.35	7.83	7124	12823	"
Miesbach, Bavaria			49.83	3.93	20.41		0.51	15.82	10.01	6136	11045	"
Feld-Bokwa Hohndorf, Saxony			82.00			12.54	0.65			7878	14180	Bunte
Zwickau, Saxony			81.58	5.74		12.68	0.93			7840	14112	"
Königshütte, Prussia			80.59	4.66	13.57	0.67	0.44			7926	14267	Schwackhöfer
Morgenstern, "			77.27	4.00	17.59	0.49	0.72			7235	13023	"
Caroline, "			77.50	4.07	17.20	0.29	0.47			7260	13068	"
Waterloo, "			79.38	4.26	15.48	0.44	0.58			7482	13468	"
Westende, "			80.03	4.38	14.40	0.47	0.55			7500	13518	"
Fanny, "			77.39	4.45	17.23	0.35	0.45			7178	12922	"
Wildensteinsegen, Prussia			79.57	4.24	15.36	0.28	0.55			7361	13250	"
Morgenroth, "			78.85	4.10	16.17	0.41	0.78			7282	13108	"
Georg, "			78.62	4.18	16.32	0.33	0.57			7208	12974	"
Eugénie, "			79.77	4.63	14.56	0.26	0.22			7427	13368	"
Veronika, "			82.37	4.94	11.90	0.22	0.77			7778	14000	"
Chasé, "			80.58	4.32	13.93	0.40				7347	13224	"

GERMANY—*Continued.*

Name or Location.	Carbon.			Hydrogen.	Oxygen.	Nitrogen.	Sulphur.	Water.	Ash.	Heat Units of Combustible.		Authority.
	Fixed.	Volatile	Total.							Calories	B.T.U.	
Louisengluck, Prussia....			80.26	4.38	14.07	0.82	0.47			7155	12879	Schwackhöfer
Maybach, Saarbruck, "			84.80	5.54		9.66				8263	14873	Bunte
Kreuzgraben, "			84.56	5.56		8.98				8099	14578	"
Camphausen, "			85.24	5.52		9.24				7975	14355	"
Von der Heydt, "			80.51	5.38		14.18				7546	13583	"
Heinitz, II. "			83.20	5.25		11.55				8008	14414	"
" II. "	61.57	18.92	80.49	4.71		14.80				8250	14850	S.-K. and M.-D.
Louisenthal, "	56.15	20.72	76.87	4.68		18.45				8020	14436	"
" "			80.43	5.34		14.23				7620	13716	"
Duttweiler, "	62.84	20.98	83.82	4.60		11.58				8500	15300	S.-K. and M.-D.
Altenwald, "	63.15	19.59	83.14	4.73		12.51				8450	15210	"
Sulzbach, "	66.30	16.75	83.05	4.95		12.00				8400	15120	"
Von der Heydt, "	61.00	20.56	81.56	4.98		13.46				8220	14796	"
Friedrichsthal, "	58.44	20.53	78.97	4.67		16.76				8220	14796	"
Hermenegilde, Silesia....			81.61	4.79	13.17	0.20	0.23			8031	14401	Schwackhöfer
Waldenburg Gluckshilf, I....			83.47	4.66	10.92	0.22	0.73			8215	14788	"
Jaklowitz..................			83.29	4.26	11.57	0.23	0.45			8086	14565	"
Michalowitz.............			81.96	4.50	12.95	0.30	0.29			7853	14136	"
Waldenburg Gluckshilf, II..			83.28	4.63	10.83	0.35	0.91			8017	14434	"
Neurod Wengelhaus, I......			82.23	4.70	11.51	0.38	1.12			7885	14193	"
" II.........			79.13	4.40	8.85	0.38	2.30			7844	14119	"
Ferdinand..................			81.40	4.65	14.29	0.32	0.34			7557	13603	"
Altendorf.................	83.87	6.05	89.92	4.18	3.90	1.00	1.00			8875	15975	S.-K.
Consolidation.............			84.29	5.27	9.37		1.07			8177	14719	Bunte
Pluto.....................			84.60	5.28	9.70		0.42			8090	14562	"
Ewald.....................			83.10	5.38	10.86		0.66			7957	14323	"
Harpener..................			87.73	4.80	6.23		1.24			8380	15084	"

GERMANY—Continued.

Name or Location.	Carbon. Fixed.	Carbon. Volatile.	Carbon. Total.	Hydrogen.	Oxygen.	Nitrogen.	Sulphur.	Water.	Ash.	Heat Units of Combustible. Calories	Heat Units of Combustible. B.T.U.	Authority.
General Erbstollen, Ruhr......	75.78	15.51		4.21	5.48			1.34	7.37	8804	15847	Bunte
Bonifazius-zeche (locomot. coal)	72.96	21.75		4.85	6.46			1.01	4.28	8411	15140	"
" " (gas)......	66.80	28.02		9.84	3.45			1.81	3.37	8028	14471	"
König Grube, Saar.............	56.52	30.67		4.67	10.98			5.09	7.72	7914	14245	"
" "	61.08	29.55		4.92	9.26			4.07	5.30	8039	14421	"
St. Ingebert (1st quality).....	64.05	30.66		5.11	8.40			2.72	2.57	8400	15120	"
" " (2d ").....	60.86	27.51		4.58	7.43			3.41	8.22	8221	14798	"
" " (3d ").....	59.76	25.66		4.60	7.97			3.12	11.44	8269	14883	"
Mittelbexbach (No. 3 seam)....	54.44	33.56		4.96	10.16			3.26	8.74	8083	14550	"
" " (No. 6 ")....	57.62	32.74		5.07	9.92			2.75	6.89	7995	14390	"
" " (No. 9 ")....	57.54	30.76		4.77	9.63			2.79	8.91	8164	14696	"
" " (No. 10 ")....	57.00	29.99		4.69	9.91			2.91	10.10	8224	14803	"
Heinitz-Dechen................	58.28	31.45		5.15	8.94			2.54	7.73	8298	14937	"
Reden-Merchweiler.............	60.08	32.20		4.98	11.53			4.80	2.92	7630	13734	"
Dudweiler.....................	65.47	28.23		5.01	7.71			2.25	4.05	8325	14986	"
Friedrichsthal................	57.01	31.23		4.74	10.40			2.99	8.77	8605	14375	"
Ziehwald......................	56.53	33.04		4.55	11.15			5.06	5.37	7511	13520	"
Louisenthal...................	58.78	30.20		4.42	12.43			6.97	4.05	7505	13509	"
Griesborn.....................	50.24	30.31		4.08	14.10			6.21	13.24	7542	13575	"
Mantau, Bohemia...............	57.76	27.23		4.46	8.86			8.78	6.23	7819	14073	"
Fremosna (Stark's)............	54.06	27.11		4.08	9.97			12.50	6.33	7468	13443	"
Turn and Taxis (large)........	58.43	29.42		4.87	10.33			7.38	4.77	7777	13998	"
" " " (average)......	57.85	29.89		4.93	9.90			6.59	5.67	7601	13682	"
Miröschau (large).............	53.80	31.01		4.58	10.97			8.84	6.35	7527	13550	"
" " (small).............	51.33	31.13		4.20	10.99			9.11	9.43	7455	13419	"
Zwickau, Saxony...............	50.91	28.22		4.18	10.34			11.16	9.71	7997	14394	"
Louisa pit, Heinitz seam, Silesia	64.30	26.76		4.62	8.92			2.83	6.11	7794	14030	"
" " Schuckman seam.....	64.20	27.75		4.74	9.82			3.25	4.80	8296	14932	"
King's pit, Gerhard seam......	61.74	29.17		4.68	10.79			4.08	5.01	7924	14262	"
" " Sattel seam........	61.50	27.48		4.74	10.54			3.78	7.24	7914	14245	"

For analyses of waste gases of these coals, see page 135.

SPAIN.

Name or Location.	Carbon.			Hydrogen.	Oxygen.	Nitrogen.	Sulphur.	Water.	Ash.	Heat Units of Combustible.		Authority.
	Fixed.	Volatile	Total.							Calories	B.T.U.	
Aller Coal.												
Prevenida.............	82.8	17.2							4.28	7350	13230	R. Oriol
Vicentera.............	82.7	17.3							5.10	7200	12960	"
Legalidad.............	81.4	18.6							4.15	7400	13320	"
Juanon................	79.1	20.9							4.90	7190	12942	"
Gabriela..............	77.7	22.3							4.80	7120	12816	"
Conveniencia.........	83.7	16.3							3.95	7250	13050	"
Ignacia...............	82.2	17.8							4.15	7180	12924	"
Dos Demigos..........	83.1	16.9							4.30	7230	13014	"
Esperanza............	77.2	22.7							4.89	7035	12663	"
Petrita...............	77.2	22.8							5.10	7010	12618	"
Mariano..............	81.9	18.1							4.50	7330	13194	"
Molino................	81.7	18.3							3.80	7180	10924	"
Catalanas	78.8	21.2							4.15	7110	12798	"
Castile.												
Joven Ildefonso, San Obrian...	73.91	26.09							8.00	6954	12517	"
Catalina, "	73.19	26.81							8.60	6930	12474	"
San Cabrian, "	78.42	21.58							10.80	7162	12892	"
Verdeña, La Permia........	58.16	41.84							4.40	5309	9556	"
San Felice..........	74.36	25.64							7.00	7075	12735	"
Villanueva, La Peña.....	88.36	11.64							2.08	7615	13707	"
Villaverde (average of 6).....	90.35	9.65							12.32	7281	13105	"
Transpeña............	91.90	8.10							3.64	7452	13413	"
Valdelora, Carrion.....	89.43	10.57							5.40	7529	13552	"

FUEL TABLES.

SPAIN—*Continued*.

Name or Location.	Carbon.			Hydrogen.	Oxygen.	Nitrogen.	Sulphur.	Water.	Ash.	Heat Units of Combustible.		Authority.
	Fixed.	Volatile.	Total.							Calories	B.T.U.	
Castile—(continued).												
Valdecastro	92.95	7.05							3.60	7705	13969	R. Oriol
Matalacasillo	84.88	15.72							8.90	7476	13457	"
Valdecorcas	87.00	12.99							2.06	6780	12604	"
Censol Menor	80.00	20.00							6.00	7395	13211	"
Valdespiña, Cea River	92.10	7.90							3.80	7796	14033	"
Tarancilla	90.72	9.28							0.84	7049	12692	"
Mosquitera, Asturia	60.05	39.95							3.60	6176	11117	"
Sama	62.40	37.60							2.50	6345	11421	"
Maria Luisa	64.40	35.60							2.55	6738	12128	"
La Jurter	68.00	32.00							2.40	7214	12985	"
Santa Barbara	76.11	23.89							3.12	7320	13176	"
Coto de Aller	83.20	10.80							5.14	7841	14113	"

CHINA.

Anthracite, Tonkin			92.45	3.06	3.47					8532	15358	Mahler

NEW CALEDONIA.

Noumea	82.75	6.25							11.00	7199	12958	L. Peletan
"	62.20	32.00			6.79				5.30	6159	11086	"
Monedon	74.23	26.00							1.77	6842	12326	"
"	86.50	6.50							7.00	7037	12667	"
Vok	78.25	9.00	83.17	10.04					12.75	6885	12393	"
Boghead, Australia										9134	16441	Bunte

RUSSIA.

Name or Location.	Carbon.			Hydrogen.	Oxygen.	Nitrogen.	Sulphur.	Water.	Ash.	Heat Units of Combustible.		Authority.
	Fixed.	Volatile.	Total.							Calories	B.T.U.	
Gangul.............			78.27	4.46	16.31	0.96				7302	13144	Alexeieff
Tkwebuli, Caucasus.....			78.42	5.13	15.41	1.04				7525	13245	"
Sosna, Altai.......			78.90	5.61	13.05	2.44				7600	13680	"
Werckne Gubach, Ural......			82.56	5.44	11.04	0.96				8116	14609	"
Rutschenkowo, Donetz.....			83.23	5.01	10.06	1.70				8230	14814	"
Kamensk Wks........			90.28	4.90	3.82	1.00				8200	14760	"
Kirghis Steppes.......			60.42	4.00	35.15	0.43				4860	8748	"
" "			62.79	5.56	31.39	0.85				5875	10515	S.-K. and M.-D.
Groucheski.........	94.90	1.76	96.66	1.35		1.99				8060	14508	"
Miouki............	81.70	9.75	91.45	4.50		4.05				8500	15300	"
Galoubosski........	63.36	19.31	82.67	5.07		12.26				7800	14040	"
Rutschenkvo........			83.43	4.98	10.82	0.77				8190	14742	Alexejew
Magatoch, Saghalien......		47.13	71.51	5.51	15.25		0.78	4.74	2.05	7888	14198	Miklaschewski
" "		41.55	65.78	5.43	14.20		0.97	5.85	7.77	7615	13707	"
Nagassi, "		29.27	69.22	5.06	8.45		0.68	1.17	15.42	8609	15296	"
Sutschan, "		7.85	86.75	4.44	1.59		0.93	0.73	5.56	8703	15665	"
Saghalien.........												

NEW ZEALAND.

Name or Location.	Carbon.			Hydrogen.	Oxygen.	Nitrogen.	Sulphur.	Water.	Ash.	Heat Units of Combustible.		Authority.
	Fixed.	Volatile.	Total.							Calories	B.T.U	
Brown coal, Malvern, Canterbury	53.29	32.04						12.65	2.02	7650	13870	Journal Iron and Steel Inst.
" " "	49.99	35.42						11.79	2.80	7338	13208	"
" " "	47.70	30.92						19.20	2.20	7444	13379	"
" Rokaia Gorge "	59.12	21.61						24.09	4.18	8751	15752	"
" Westport	56.01	37.17						2.60	4.22	7545	13481	"
" "	57.92	34.94						3.96	3.18	7953	14315	"
" Waikato, Auckland	50.01	29.97						19.82	2.20	7948	14306	"
" Shag Point, Otago.	45.30	30.10						19.20	5.40	7250	13050	"
" Kaitangata, "	44.11	38.32						15.44	2.13	6725	12105	"
" " "	39.41	27.25						19.61	3.73	6451	11612	"
" Okoko, Auckland.	39.83	33.74						22.21	4.22	6772	12190	"
" Springfield, Colliery	38.00	31.50						18.60	11.90	6809	12256	"
Bituminous, Grey River, West'nd	62.37	29.44						1.99	6.20	8425	15165	"
" Preservation Inlet.	60.88	20.69						4.33	6.19	8536	15364	"
" Westport.......	59.75	32.14						3.97	4.14	8155	14679	"
" Mokihinui	55.59	38.86						3.16	2.39	7362	13252	"
" Brunner Mine....	56.62	35.68						1.59	6.11	7701	13862	"
" Otamatawea Creek.	52.89	36.63						2.19	8.29	8328	14990	"
" Cape Farewell....	48.59	43.17						2.18	6.06	6970	12546	"
" Ross, Southland...	42.53	31.43						6.58	19.26	5470	9846	"
Glance coal, Whangaree, Auckland........	50.11	38.68						8.01	3.20	7070	12726	"
Black coal, Kawa River	50.15	42.63						4.18	3.04	6786	12215	"
Pitch coal, Grey River, Westland	60.20	29.97						8.01	1.81	8375	15075	"
" Whangaree.......	50.01	37.69						9.61	2.69	7157	12882	"

LIGNITES.

AMERICAN.

Name or Location.	Carbon.			Hydrogen	Oxygen, Nitrogen, Sulphur.	Oxygen.	Nitrogen.	Sulphur.	Water.	Ash.	Heat Units of Combustible.		Authority.
	Fixed.	Volatile	Total.								Calories	B.T.U.	
Erie, Colorado	45.98	32.71		4.25		6.65	1.64	0.52	18.57	2.74	6311	11360	Anonymous
Black Diamond, Colorado	39.01	43.05		5.46		13.32	1.45	0.77	14.67	3.27	5837	10507	"
Cañon City, Colo., vertical	51.36	37.61		7.38		9.27	1.50	1.02	7.01	4.03	7276	13097	"
" " " upper	47.93	36.74		5.37		12.14	1.75	0.62	6.56	8.76	6206	11170	"
" " " lower	49.54	37.21		5.48		11.39	1.45	0.82	7.66	5.59	6469	11644	"
Golden City, " 12 ft. seam	38.46	41.23		4.89		13.88	0.95	0.30	17.64	2.67	5526	9947	"
" " 4 "	34.89	44.74		5.14		14.60	1.50	0.42	17.15	3.22	5432	9778	"
" " 3 "	42.08	36.20		5.07		27.77	1.20	0.43	18.35	3.37	4530	8154	"
Gunnison River, Colo.	84.65	12.16		3.72		4.20	1.65	0.70	1.50	2.29	7911	14240	"
Lechner's South Park, Colo.	58.62	33.79		5.23		12.86	2.35	0.47	6.30	1.28	6980	12204	"
Marshall, Colo.	46.43	37.84		7.43		12.10	1.34	0.66	13.19	2.54	6510	11478	"
Mt. Carbon, Colo	37.82	36.91		4.91		6.68	1.25	0.40	20.38	4.87	5902	10624	"
Elsinore, California											5935	9063	E. E. Slosson
" (briquette)											6235	11205	

EUROPEAN.

Name or Location.	Carbon.			Hydrogen	Oxygen, Nitrogen, Sulphur.	Oxygen.	Nitrogen.	Sulphur.	Water.	Ash.	Heat Units of Combustible.		Authority.
	Fixed.	Volatile	Total.								Calories	B.T.U.	
Freienstein, Austria			72.12	4.83	23.05						6472	11650	Schwachhöfer
Briex, "			71.78	4.93	23.29						6638	11950	"
Koeflach, "			67.93	5.45	26.62						6096	10973	"
Panhraz, Bohemia			82.23	4.61	13.16						7604	13687	"
Bilin, "			72.17	5.49	22.34						6364	11731	Bunte
Bustehrad-Kladno, Bohemia			79.94	4.41	15.65						7448	13407	Schwachhöfer
Fat lignite, Bohemia			76.58	8.27	15.15						7024	12263	Scheurer-Kestner
Rocher-Bleu, France			72.98	4.94	22.98						6480	11664	
Manosque, "			70.57	5.44	23.99						7363	13253	
" "			66.31	4.85	28.84						6991	12584	
Toula, Russia			73.72	6.09	20.19						7687	13637	Schwachhöfer
Salgo-Tarjan, Hungary			71.78	5.33	21.89					11.20	6244	11239	Przlwoznik
Dodoşé, Kralyrcani, Hungary			47.2	4.25		18.15			19.20	13.45	6951	12511	"
Ilz "			40.1	4.23		24.07			9.15	8.99	5372	9670	
Nagy Kavocsi			51.6	4.10		19.80			15.66	8.74	6074	10933	
Pernik, Bulgaria		35.07	60.42	4.00		35.15	0.43		12.47		5816	10472	Anonymous
Kirghis Steppes, Russia	41.91		62.79	5.56		31.39	0.20				4860	8748	Alexcieff
" "											5875	10475	"
Brown coal, Bilin, Germany	23.33	47.83		3.75		15.23		1.86	25.92	2.92	6090	10962	Bunte

PEAT.

Location.	Carbon.	Hydrogen.	Oxygen and Sulphur.	Ash.	Water.	Coke.	Calories.	B.T.U.	Authority.
Ismaning, Bohemia	39.26	3.88	22.93	7.27	26.26	28.00	3246	5843	Bunte
" " (pure)	63.13	5.88	30.99			31.42	4978	8960	Bunte
Rosenheim, "	44.08	4.54	28.34	2.00	21.04	27.4	4115	7407	Bunte
" " (pure)	57.27	5.89	36.84			33.02	5347	9625	Bunte
Zengermoos, "	40.82	4.43	26.58	7.72	20.84	?	3774	6793	Bunte
" " (pure)	59.14	6.24	36.62			?	5283	9509	Bunte
Bohemia (locality not given)	53.18	5.54	34.23	0.93	6.12	31.7	5489	9880	
" (pure)	57.21	5.96	36.83				5903	10625	
Dried peat							5940	10692	Bainbridge
Moist peat					6.12		4675	8415	Bainbridge
Bohemia	53.19	5.54	34.23	0.92			5903	10625	Mahler
Ireland, (dry)	59.00	6.00	30.00	4.00			5700	10260	Berthelot and Petit.
" (moist)							4106	7391	Berthelot and Petit.

WOOD.

Name.	Carbon.	Hydrogen.	Oxygen.	Nitrogen.	Ash.	Water.	Calories.	B.T.U.	Authority.
Ash	49.18	6.27	43.91	0.07	0.57		4711	8480	Gottlieb
Beech	49.06	6.11	44.17	0.09	0.57		4774	8591	"
Birch	48.88	6.06	44.67	0.10	0.29		4771	8586	"
Elm	48.89	6.20	44.25	0.06	0.50		4728	8510	"
Fir	50.36	5.92	43.39	0.05	0.28		5035	9063	"
Oak	50.16	6.02	43.36	0.09	0.37		4620	8316	"
Pine	50.31	6.20	43.08	0.04	0.37		5085	9153	"
Norway pine	47.37	5.58	39.78		0.34	6.94	4828	8690	Mahler
Oak from Lorraine	46.59	5.43	40.34		0.75	6.93	4689	8440	"
Pine-knot, Texas							6035	10863	Slosson
Tan bark		.			15.0	30.0	3389	6100	Peclet
"							2380	4284	"
Straw, wheat, Russia						6.20	5770	10386	Eng. Mechanics, February, 1893
" (dry)	4.1.1	5.6	43.7	0.42	4.10	0.00	6290	11322	"
" "						10.00	5448	9806	"
" buckwheat (dry)							5590	10068	"
" flax (dry)							6750	12150	"

OVEN COKES.

AMERICA.

Name or Location.	Carbon.			Hydrogen.	Oxygen.	Nitrogen.	Sulphur.	Water.	Ash.	Heat Units of Combustible.		Authority.
	Fixed.	Volatile	Total.							Calories	B.T.U.	
Connellsville, Pa............	89.58	0.46					0.81	0.03	9.11	7895	14211	Anon
Coketon, Pa.................	86.58	1.26					1.49		10.67	8015	14427	"
Beaver Falls, Pa............	84.73	0.63					1.99	0.01	12.64	8039	14471	"
Clarion Co., Pa.............	88.36	1.11					1.08	0.23	9.23	8022	14440	"
Bloss, Pa...................	81.92	0.57					0.90	0.78	7.95	7983	14369	"
Coalbury, Pa................	84.67	0.68					1.87	0.13	12.63	7897	14210	McCreath, analyst
Connellsville, Pa............	89.51	0.88					0.71	0.70	8.83	7911	12240	Wedding
" (average of 3).	87.46	0.01					0.69	0.49	11.32	8070	14526	Morrell
"	88.96						0.81		9.74	8020	14436	"
Dade, Pa....................	75.94	0.09					0.67	0.54	21.75	7953	14315	McCreath
Daisy, Pa...................	79.83	1.05					2.13	1.22	15.75	7731	13883	"
Etna, Pa....................	85.45	1.16					1.45	0.86	11.08	7896	14212	"
Fairmont, Pa................	85.78	0.62					2.11	0.30	11.46	8034	14464	Anon
Irwin, Pa...................	88.24	1.38					0.96		9.41	8026	14447	"
Jefferson Co., Pa............	88.95	1.42					0.90	0.78	7.95	7983	14369	"
Kittanning, Pa..............	87.22						1.23		11.43	8046	14483	"
Pratt, Pa...................	88.87	1.58					1.18	1.92	8.99	7946	14300	McCreath
St. Bernard No. 9, Pa.......	86.48	1.13					1.05	0.13	11.21	7937	14286	"
" No. 11, Pa......	87.27	0.61					2.21		12.11	8075	14543	Morrell
Seymour, Pa.................	90.69	0.34					2.37		8.96	7995	14340	"
Snow Shoe Co., Pa...........	90.65	0.63					0.85	0.22	7.65	8036	14468	Anon
	82.63	2.95					1.10	0.99	12.33	8017	14431	"

AMERICA—*Continued.*

Name or Location.	Carbon.			Hydrogen.	Oxygen.	Nitrogen.	Sulphur.	Water.	Ash.	Heat Units of Combustible.		Authority.
	Fixed.	Volatile	Total.							Calories	B.T.U.	
Fairmont, W. Va............	91.08	1.85					0.67	0.24	6.83	8025	14445	Wedding
Flat Top, W. Va............	92.55	0.76					0.60	0.35	5.75	8011	14420	"
New River, W. Va. (av. of 8)..	92.38						0.56	1.11	7.21	8015	14427	Proctor
Pocahontas, " "	91.68	0.52					0.75		7.57	8090	14562	McCreath
" (av. of 3)..	92.53						0.60		5.74	7988	14378	Proctor
" "	91.43	1.27					0.51	0.70	6.09	8006	14411	Anon
" "	92.82	0.66					0.55	0.66	4.91	8032	14457	"
" "	92.55	0.76					0.60	0.35	5.75	8011	14418	"
" "	92.58	0.49					0.68	0.29	6.05	8000	14400	"
Pineville, W. Va............	94.66	0.04					0.69	1.14	3.57	8006	14128	Morrell
Big Stone Gap, Ky. (av. of 7)..	93.23						0.75		5.69	8044	14479	Proctor
De Bardeleben, Ala. (48 hours)	86.00	0.70					1.23	0.40	12.90	8019	14435	Phillips
" " (72 hours).	87.25						0.99	0.05	11.80	7998	14397	"
Birmingham, Ala. (av. of 4)...	87.29	0.90					1.19		10.54	7940	14290	Proctor
Chattanooga, Tenn. (av. of 4)..	80.51						1.59		16.34	7832	14097	"

FRANCE.

Name or Location.	Carbon.			Hydrogen.	Oxygen.	Nitrogen.	Sulphur.	Water.	Ash.	Heat Units of Combustible.		Authority.
	Fixed.	Volatile	Total.							Calories	B.T.U.	
Anthracite..................			91.04	0.68	2.15			0.23	5.90	8036	14465	Mahler
Antin......................			91.58	0.63	1.58				3.20	8004	14407	"
Commentry.................			92.73	0.44	2.63				4.20	8001	14402	"
Grand Combe			89.27	0.21				0.59	7.80	7920	14256	"
Mons			92.34	0.34	2.22					7936	14284	De Marcilly

OVEN COKES.

ENGLAND.

Name or Location.	Carbon.		Hydrogen.	Oxygen.	Nitrogen.	Sulphur	Water	Ash.	Heat Units of Combustible.		Authority.
	Fixed.	Volatile							Calories	B.T.U.	
	Total.										
Best, Durham................			93.15	0.72	0.90	1.28	0.65		3.95	8101	14582	Kubale
Average, Durham............			84.92	4.53	6.66	1.96	0.81		2.28	7899	14218	"
Busty, Durham (upper layer)...			81.22	4.70	9.45		1.83	0.85	3.28	7911	14240	"
" " (lower layer)..			78.46	4.42	8.82		1.00	0.99	6.17	7959	14326	"
Brockwell, Durham...........			83.40	4.40	7.18		0.81	0.99	3.50	7811	14060	"
Hamsteels...................			92.55				0.84	0.21	6.36	8011	14420	"
Consett.....................			91.88				1.21	0.37	6.91	8020	14436	"
Whiteworth.................			91.56					0.54	6.69	8002	14404	"
South Braucepeth............			93.43				0.91	0.36	5.30	8005	14409	"

NEW SOUTH WALES.

Mount Pleasant..............	85.85	0.50						0.20	13.45	8043	14477	Murgaye

GAS COKES.

GERMANY.

Name or Location.	Carbon.		Hydrogen.	Oxygen and Nitrogen.	Sulphur.	Water.	Ash.	Heat Units Combustible.		Authority.
	Fixed.	Volatile						Calories	B.T.U.	
		Total.								
Sulkov, Bohemia..........	87.23	1.76	0.53	2.96	1.39	0.90	9.62	7865	14157	Bunte
Pankraz, Bohemia.........	83.74	2.47	0.83	3.84	2.31	0.81	11.48	7790	14022	"
Consolidation, Ruhr.......	90.58	0.21	0.70	4.04	1.79	0.87	7.42	7785	13913	"
Rhone-Elbe and Alma, Ruhr.	89.75	2.04	0.81	4.80	1.71	0.88	6.50	7716	13889	"
Ewald, Ruhr.............	83.98	2.51	0.90	3.74	2.53	1.17	11.18	7781	14006	"
Bonifacius, Ruhr..........	84.56	3.17	1.07	3.61	1.53	1.02	10.74	7819	14074	"
Camphausen, Saar.........	85.14	2.80	1.00	2.60	1.79	1.43	10.27	7900	14226	"
Heinitz	91.78	0.74	0.78	2.85	0.96	0.81	6.52	7862	14152	"
United Glückhilf, Lower Silesia.	85.51	1.34	0.92	2.34	1.52	1.24	11.60	8022	14440	"
Deutschland, Upper Silesia..	90.62	2.46	0.75	3.13	3.20	0.96	3.72	7826	14087	"
Königin Louise, Upper Silesia.	87.93	0.93	0.54	2.01	0.96	3.73	6.41	7938	14288	"
Saarbruck...............	98.04	0.73						8200	14760	"
Petroleumcoke	98.05	0.50						8057	14503	Mahler
Bohemian coke	98.81	1.19						8290	14922	Bunte

OILS.
AMERICA.

Name or Location.	Sp. Gr.	Carbon.	Hydrog'n	O + N	Oxygen.	Nitrogen.	Calories.	B.T.U.	Authority.
Heavy petroleum, W. Virginia.	0.873	83.5	13.3	3.2			10180	18324	St.-Claire Deville
Light petroleum, " "	0.841	84.3	14.1	1.6			10223	18400	"
" " Pennsylvania.	0.826	82.0	14.8	3.2			9963	17930	"
Heavy petroleum, "	0.886	84.9	13.7	1.4			10672	19210	"
" " Ohio.	0.887	84.2	13.1	2.7			10399	18718	"
American oil sold in Paris.		83.4	14.7	1.9			9771	17590	
" " heavy, "		86.9	13.1				10913	19643	Mahler
" " refined, "		85.5	14.2	0.3			11045	19881	"
" naphtha, "		80.6	15.1	4.3			11086	19955	"
" crude, "		83.0	13.9	3.1			11094	19970	"
Heavy Pennsylvania oil.	0.886	84.9	13.7		1.4		10680	19224	Robinson
" W. Virginia oil.	0.928	88.3	13.9		0.8		10102	18184	"
No Wood, Wyoming.	0.9960						10927	19668	C. E. Slosson and
Shoshone Reservation, Wyoming							10883	19590	L. C. Colburn
Salt Creek, Natrona Co., "							10813	19463	"
Oil Mt., Natrona Co., "							10743	19337	"
Newcastle, Western Co., "							10447	18805	"
Little Popo, Agie, "	0.900—.921						10430	18774	"
Landen.	0.8565—0.8635						10883	19590	"
Oil Creek, Pa.	.730	82.0	14.8		3.2	0.54	11606	20890	Anon
Scotia well, W. Va.		86.6	12.9				11801	21240	"
Cumberland, W. Va.		85.2	13.4				11483	20670	"
Bothwell, Canada.	0.857	84.3	13.4		2.3		11339	20410	"
Petrolea, "	0.870	84.5	13.5		2.0		11406	20530	"
Residuum, Virginia oil	0.860	87.1	11.7		1.2		10667	19200	"
Pennsylvania crude	0.938	84.9	13.7		1.4		11520	20736	"
California, Hayward Company.		86.9	11.8			1.1	11728	21110	"
Lima, Ohio.		80.2	17.1	2.7			12000	21600	Mayer
Mineral seal.	0.83	83.3	13.2				11147	20065	Stillman & Jacobus

FOREIGN.

Name or Location.	Sp. Gr.	Carbon.	Hydrog'n	O + N	Oxygen	Nitrogen.	Sulphur.	Calories.	B.T.U.	Authority.
Petroleum.										
Salo, Parma	0.786	84.0	13.4	1.8				10120	18216	St.-Claire Devile
Pechelbronn	0.912	86.9	11.8		1.3			9708	17475	"
Crude, Pechelbronn	0.892	85.6	9.6		4.5			10020	18036	"
Schwabwiller, Lower Rhine	0.861	86.2	13.3	0.5				10458	18781	"
Gallicia, East	0.872	82.2	12.1	5.7				10005	18010	"
" West	0.885	85.3	12.6	2.1				10231	18416	"
Baku (heavy)		87.0	12.9					10843	19517	Mahler
Novorossik		84.9	11.6	9.5				10328	18590	"
Crude Naphtha, Balchany	0.884	87.4	12.5	0.1				?11070	19926	"
Residuum, Balchany	0.928	87.1	11.7	1.2				?11700	21060	"
Black oil, Weyser works		86.5	12.0	1.5				?10760	19368	"
Light oil, Baku	0.884	86.3	13.6	0.1				11460	19700	"
Heavy oil, "	0.956	86.6	12.3	1.1				10800	19440	"
Timaacon, Java	0.923	87.1	12.0	0.9	0.9	0.25		10831	19496	"
Madja, "		83.6	14.0	2.4				9593	17267	"
Kendong, "		85.0	11.2	2.8				10183	18330	"
Ograio	0.985	87.1	10.4		2.5				18145	Anon
Paraffin oil (pressed)								10172	18310	"
Russia, crude	0.884	86.3	13.6		0.1			12650	22628	Goulishambaroff
" " "	0.938	86.6	12.3		1.1			10800	19440	"
Baku, Russia, crude	0.938							11200	20160	"
" " refuse	0.928	87.1	11.7		1.2			10700	19200	"
" " crude								11760	21168	"
" " astatki	0.928	84.49	13.96		1.25			10340	18611	W. Thompson

OILS.
FOREIGN—Continued.

Name or Location.	Sp. Gr.	Carbon.	Hydrog'n	O + N	Oxygen.	Nitrogen.	Sulphur.	Calories.	B.T.U.	Authority.
Schist oil.										
Crude from Vaquas, Ardèche..	0.911	80.3	11.5	8.2				9046	16282	St.-Claire Deville
Autun		79.7	11.8	8.6				9950	17910	"
Nouvelle Galles		83.7	11.8	4.4				10381	18686	Mahler
Heavy								9654	17377	St.-Claire Deville
Crude								8623	15520	"
Acid residuum								9836	17705	"
From la Condamine								9612	17300	"
Gas oil from Paris Works	1.044	82.0	7.6	10.4				8916	16050	"
Ozokerite, Boryslaw		85.2	14.7	0.11				11163	20093	Mahler
Ozokerite oil								10648	19122	St.-Claire Deville
Heavy pine oil (blue, fluorescent)								10081	18146	"
Blast furnace oil, Coatbridge, Eng		84.8	10.5	4-7				8933	16080	W. Thompson
Olive oil (pressed)		83.64	10.59		5.94		0.087	9328	16790	Stohmann
"								9473	17050	"
Poppy seed oil (pressed)								9442	16996	"
Rape seed oil (pressed)								9489	17080	"
" " "								9619	17314	"
Sperm oil								10000	18000	Gibson

ASPHALT AND BITUMEN.

Name or Location.	Fixed.	Volatile	Total.	Hydrog'n	Oxygen.	Nitrogen.	Sulphur.	Ash.	Calories	B.T.U.	Authority.
Bitumen, pure	24.80	72.44	77.68	8.02	14.30			2.76	8408	15134	Johnson
Boghead shale, Australia			83.17	10.04	6.15				9134	16431	Bunte
Asphalt, Dead Sea			76.36	9.16	11.53	0.46	3.00	3.62	8900	16020	Mahler
" Big Horn, Wyoming									9532	17159	Slosson
" Walled Creek, "									6307	11953	"

NATURAL GAS.

Name or Location	Hydrogen.	Methane, CH_4.	Ethylene, C_2H_4.	Illuminants.	CO_2	CO	Oxygen.	Nitrogen.	H_2S	Calories per Cubic Meter.	B.T.U. per Cubic Foot.	Authority.
Anderson, Indiana	1.86	93.07	0.47		0.26	0.73	0.42	3.02	0.15	9494	1021	E. & M. Journal
Kokomo, "	1.42	94.16	0.30		0.27	0.55	0.30	2.80	0.18	9581	1030	"
Marion, "	1.20	93.57	0.15		0.20	0.60	0.55	3.42	0.20	9518	1024	"
Muncie, "	2.35	92.67	0.25		0.25	0.45	0.35	3.53	0.15	9477	1019	"
Louisville, Kentucky	1.31	87.75			6.60			4.34		8849	939	Slocum
Olean, New York		96.50		1.00		0.50	2.00			9900	1071	R. Young
W. Bloomfield, New York		82.41		2.94	10.11		0.23	4.31		9158	998	H. Wurtz
Findlay, Ohio	2.18	92.60		0.31		0.50	0.34	3.61	0.20	10250	1100	E. McMillin
"	1.64	93.35	0.35		0.25	0.41	0.39	3.41	0.20	9486	1020	E. & M. Journal
Fostoria, "	1.89	92.84	0.20		0.20	0.55	0.35	3.82	0.15	9450	1016	"
St. Mary's, "	1.94	93.85	0.20		0.23	0.44	0.35	2.96	0.21	9523	1028	"
Burn's well, St. Joe, Penna	6.10	75.44	18.12	trace	0.34	trace				10090	1170	S. P. Sadtler
Cherry Tree, Penna	22.50	60.27	6.80		2.28		0.38	7.32		8034	840	"
Creighton, "		96.34		trace	3.64					9671	1025	F. C. Phillips
E. Liberty, "	9.64	57.85	0.80	5.20		1.00	2.10	23.41		5581	592	S. A. Ford
Harvey well, "	13.50	80.11	5.72	0.56	0.66	0.26				9331	990	S. P. Sadtler
Leechburg, "	4.89	89.65	4.39	28.87	0.35	0.22	0.16	27.87		9962	1073	"
Grapeville (dry), Penna	7.05	35.08	0.17	39.64	0.58	trace	0.12	18.69		7698	823	Morrell
"	24.56	14.93	0.96		trace	trace	2.20			8326	891	"
Murraysville, Penna	19.56	78.24		6.30	0.80	1.00	0.80			8458	900	Rogers
Pittsburg, "	20.02	72.18		4.8	3.6			8.9		8620	917	Anon
Pechelbronn, Germany		77.03		4.26	3.50	3.5	1.8			8539	908	"
Caspian Sea, Russia		92.24		4.11	0.93					9859	1062	"
Aspharon Peninsula, Russia	0.34	92.49			0.60			2.13		9859	1062	"
Blower in Mine, Wales		95.42						3.78		9578	1014	J. W. Thomas
Occluded gas, Wigan, Wales		80.69	4.75		6.44			8.12		8770	930	"

NATURAL GAS (FIRE DAMP).
ENGLAND.

Name or Location.	Sp. Gr.	Hydrogen.	Methane.	Illuminants.	Carbonic Acid.	Oxygen.	Nitrogen.	Air.	Calories per Cubic Meter.	B.T.U. per Cubic Foot.	Authority.
Bensham seam, Wallsend	.6024		91.00		1.30			9.00	9134	976	Turner
" " "			77.50		0.30	0.60	21.10		7779	832	Playfair
Pipe from " "			92.80		0.70	0.90	6.90		9315	996	"
Below Bensham, Hebburn			91.80		0.90		6.70		9215	985	"
" " (30 d. later)			92.70		1.60		6.40		9304	995	"
Bensham, "			86.50				11.90		8682	928	"
" Jarrow	.6381		81.50		2.10	0.40	14.20	18.50	8132	869	Turner
" "			83.10		1.70		4.90		8331	890	Playfair
Five quarter, "			93.40			3.00	12.30		9375	993	"
Low Main, "		3.00	79.70						8090	865	"
Same (deeper)	.6209		89.00					11.00	8934	955	Turner
Yard Coal, Burraton	.6000		91.00				7.00	9.00	9130	976	"
High Main, Killingworth	.6196		85.00				16.50	8.00	8530	912	"
Low " "	.8226		37.00			1.00	16.50	46.50	1317	397	Graham
" " "	.6306		82.50				6.32		8280	885	Richardson
Main 100 fathoms, Hutton			66.30		4.03		27.00	23.35	6652	711	Turner
Hutton 175 " "	.7800		50.00				44.00	23.00	5010	535	"
" waste 125 fathoms, Pensher	.7470		50.00				11.00	6.00	5019	536	"
Three-quarter, Townley	.9660		7.00		6.00		4.68	82.00	703	75	Richardson
Two-quarter, Well-gate			56.17		0.50	1.30	1.30	33.15	5638	602	Playfair
" Gates-head	.5802		98.20		0.80	15.50	4.50		9857	1044	Graham
Cwm-Twrch			94.20				63.80		9457	1002	Graham
			19.30						1937	206	Playfair

COAL GAS.

AMERICA.

Name or Location.	Hydrogen.	Methane, CH_4.	Ethylene, C_2H_4.	Illuminants.	Carbonic Acid.	Carbonic Oxide.	Oxygen.	Nitrogen.	Water.	Calories, per Cu. Meter.	B.T.U per Cu. Foot.	Authority.
From Dominion N. S. coal	50.4	33.7		5.2	2.6	5.8	0.8	1.5		5775	618	Slocum
" Westmoreland coal	40.9	42.8		5.9	1.8	3.6	1.1	3.9		6440	690	"
Coal (average)	48.1	36.5		4.3	0.3	7.6	0.4	2.8		5917	632	Arion.
"	46.00	40.0		4.0	0.50	6.0	0.5	1.5	1.50	6882	735	E. McMillin
Coke ovens, Johnston, Pa.	57.2	18.8		0.8	2.00	3.20		18.0	H_2S	3736	399	Dr. W. E. Rothberg
" Westphalia	53.2	36.11		2.24	1.41	6.49			0.43	5730	612	Otto Hoffman
Newton, Mass.	50.59	34.80		5.23	1.16	6.16		2.06		5608	599	C. D. Jenkins
Cambridge, Mass				C_2H_6						5393	576	"
Cleveland, O.	34.80	28.80	9.50	1.70	0.20	10.40	0.40	14.20		6151	657	Thwaite
Hoboken, N. J.	39.50	37.30	5.85	0.75	2.70	4.30	1.40	8.20		6039	645	"
Boston, Mass.	47.49	38.67	5.21		1.04	6.74		0.85		6095	651	"
Cincinnati, O.	45.85	39.26	5.17		0.82	4.78	0.41	3.71		6039	645	"
Cape Breton, Canada	44.6	39.2		6.2	1.4	4.5	0.6	3.3		5460	612	F. L. Slocum
" "	45.4	36.5		5.2	2.2	3.6	0.6	6.3		5455	611	"
International, Canada	46.5	35.7		5.0	3.1	5.7	0.5	3.7		5536	620	"

EUROPE.

Name or Location.	Hydrogen.	Methane, CH_4.	Ethylene, C_2H_4.	Illuminants.	Carbonic Acid.	Carbonic Oxide.	Oxygen.	Nitrogen.	Water.	Calories, per Cu. Meter.	B.T.U per Cu. Foot.	Authority.
Birmingham, Eng.	40.23	39.00		4.76	1.50	4.05	0.36	10.10		6283	671	Thwaite
Coke ovens, Brymbo	45.9	22.9		3.7	3.3	9.9		13.5	0.8	5234	559	Chas. Hunt
Bonn, Germany	39.80	43.12		4.75	3.02	4.66		4.65		6666	712	Thwaite
Brighton, Eng.	51.62	38.15		3.76	0.03	4.14	0.23	2.07		6320	675	"
Bristol, Eng.	44.57	40.70		4.58		4.77	0.27	5.11		6563	701	"
Chemnitz, Germany	51.29	36.45		4.91	1.08	4.45		1.41		6395	683	"

244 FUEL TABLES.

EUROPE—Continued.

Name or Location.	Hydrogen.	Methane CH$_4$.	Ethylene C$_2$H$_4$.	Illuminants.	Carbonic Acid.	Carbonic Oxide.	Oxygen.	Nitrogen.	Sulphuretted Hydrogen.	Calories, per Cu. Meter.	B.T.U. per Cubic Foot.	Authority.
Cologne, Germany	56.90	34.40		3.50	1.50	5.20	1.40	4.00		6095	651	Thwaite
Dresden, "	48.70	33.40		3.00	1.00	8.00	1.00	3.64		5711	610	"
Edinburgh, Scotland	33.24	42.93		12.23	0.35	6.61	0.06	3.07		8370	894	"
Glasgow, "	39.18	40.26		10.00	0.29	7.14	0.06	2.73		7774	830	"
Gloucester, England	48.89	38.25		4.95	0.03	4.64	9.51	1.01		6563	701	"
Hanover, Germany	46.27	37.55		3.17	0.81	11.19		4.23		6188	661	"
Heidelberg, "	44.00	38.40		7.27	0.37	5.75	0.12	10.84		6797	726	"
Ipswich, England	43.23	38.73		4.53	0.06	2.46	0.07	4.32		6272	670	"
Leeds, "	40.26	42.74		7.28	0.34	5.02	0.19	6.10		7247	774	"
Liverpool, "	36.44	41.28		7.90	1.70	3.39	0.26	5.95		7415	792	"
London G. L. & C. Co., Eng.	47.99	37.64		4.41		3.75	0.1	4.3		6320	675	"
" " " "	49.2	34.2		4.7		7.5		3.19		6114	653	Chas. Hunt
So. Metropolitan, Eng.	53.14	36.55		2.92	0.09	4.11	0.11	3.19		6030	644	Thwaite
Manchester, England	45.58	34.90		6.46	3.67	6.64		2.46		6535	698	"
Newcastle-under-Lyme, Eng.	46.31	39.01		4.53	0.08	3.74	0.11	6.22		6395	683	"
Newcastle-on-Tyne	50.50	36.71		3.62	0.28	3.37	0.23	5.29		6095	651	"
Norwich, "	53.79	36.11		3.26	0.27	3.40	0.14	3.03		6049	646	"
Nottingham, "	45.52	39.66		5.63	0.81	5.63	0.24	2.51		6797	726	"
Paris, France	50.10	33.10		5.80	1.50	6.30	0.50	2.70		6244	667	"
Preston, England	43.95	39.33		6.22	0.84	4.62	0.25	4.79		6825	729	"
Redhill, "	48.18	39.41		4.40	0.74	3.41	0.49	3.37		6498	694	"
St. Andrews, Scotland	36.63	42.13		10.04	2.73	5.16	0.48	2.83		7827	836	"
Sheffield, England	43.05	43.05		6.28	0.24	4.72	0.10	2.56		7163	765	"
Southampton, "	53.59	38.15		3.76	0.03	4.14	0.23	2.07		6413	685	"
Oil Gas	46.2	37.5		4.9		6.8		4.6		5768	616	Chas. Hunt
" "	31.61	46.17	C$_2$H$_2$ 16.29			0.14		5.06		8124	869	Rogers
" Pintsch	18.4	42.9	11.3	15.5			0.73	11.8		12660	1320	Worstall and Burwell

AIR AND WATER GAS.
AMERICA.

Name or Location.	Hydrogen.	Methane.	C_2H_4	Illuminants.	CO_2	CO	Oxygen.	Nitrogen.	H_2S	Calories, per Cubic Meter.	B.T.U. per Cubic Foot.	Authority.
Anthracite gas	52.76			4.11	2.05	35.38		4.43		3385	385	Von Langen
American coal	45.0			2.00	4.00	45.00		2.00		2816	301	Taylor
Coal and Granger, Charlestown, Mass.	42.80	28.63		10.42	2.19	13.93		2.03		5738	614	Jenkins
Coke	50.10			0.70	4.00	40.00		5.3		2859	294	Von Langen
Coke and bituminous coal	94.08				0.50	3.54	0.13	0.12		3032	324	"
Evans process, Jackson, Mich.	52.39	8.51		0.5	0.62	36.31		1.54		3757	402	C. H. Evans
Fahnejelm process, Chicago, Ill.	50.9	3.2			6.0	32.2	7.7			3014	322	D. Fisher
Flannery proc. So. Boston, Mass.										5934	635	Jenkins
Granger process (uncarburetted)	52.88	2.16		3.47		36.8		4.69		2642	283	G. E. Moore
" " (carburetted)	30.0	24.0	12.5	0.3	4.8	29.0	0.2	2.5		6000	640	"
" " (from coke)	52.41	0.2		0.9	8.2	11.5	0.6	0.47		3098	331	Slocum
Hastings " (from Jellico coal)	43.8	8.2		1.4	7.9	32.8	0.1	5.2		3394	363	"
" " (from slack)	48.4	8.2		4.5	10.5	28.5	1.0	5.2	1.5	3642	390	" [& Schimmel
Hemnin "	38.0			0.10	8.14	20.0	0.70	26.0		2720	290	Shepard, Bruckner
Loomis " Akron, Ohio.	46.54	2.51		0.29	7.60	31.56		10.45		2626	281	R. Norris
" " Boston, Mass.	53.40	3.10		0.34	6.02	29.50		6.05		2884	308	E. C. Jones
" " Bridgeport, Conn.	49.68	3.46		0.04	7.10	30.28	0.53	10.22		2809	300	E. G. Love
" " Tacony, Pa.	50.70	3.20		1.20	4.05	31.96		6.11		2841	307	R. Norris
" "	55.0	5.75				31.00		3.00		3187	341	H. E. Loomis
Lowe process, Des Moines (with 1¾ gal. oil)	41.7	12.2		5.4	4.5	34.6	0.4	1.2		4580	490	W. W. Randolph
Lowe process, Des Moines (with 2⅞ gal. oil)	37.6	16.5		8.9	3.7	30.7	0.7	1.9		5514	590	

246 FUEL TABLES.

AIR AND WATER GAS.
AMERICA—Continued.

Name or Location.	Hydrogen.	Methane.	C_2H_4	Illuminants.	CO_2	CO	Oxygen.	Nitrogen.	C_6H_6	Calories, per Cubic Meter.	B.T.U., per Cubic Foot.	Authority.
Lowe process, Jersey City, N. J.	37.95	17.83	2.8	8.0	2.2	24.2		5.02	2.00	6903	739	Shepard, Bruckner & Schimmel
" " Long Island City, N.Y.	34.0	20.8	8.0	16.6	1.6	23.0	0.4	5.2	2.00	7183	768	"
Lowe process, Long Branch, N.Y.	31.3	19.1		13.6	3.0	27.4	0.4	3.8	1.4	6878	736	F. B. Wheeler
Lowe process, Philadelphia, Pa.	50.9					44.5	0.07	2.08		3062	327	E. E. Taylor, 1885
" " Ex-position, Pa.	44.5				3.6	42.1		9.8		2458	263	Dr. Greene
Lowe process, (Oil-water)	29.3	22.1		13.9	4.0	25.8	0.7	4.2		6295	705	F. L. Slocum
" " (Blue Gas)	52.4	0.8			4.6	41.5		0.5		2964	332	"
New York Gas Exposition, N. Y. City, 1897	32.7	16.8		14.4	2.4	30.2	0.4	3.1		7160	766	F. B. Wheeler
Rose-Hastings, Louisville, Ky. (from soft coal)	36.4	23.2		14.05	3.02	19.1	1.15	3.08		6140	657	W. A. Noyes
Rose-Hastings (generator gas)	9.8	49.6		1.1	8.1	28.1	0.3	3.9		3482	390	F. L. Slocum
" (enriched)	26.0	34.6		11.9	5.6	10.9	0.3	1.6		6000	673	"
Strong process, Mt. Vernon, N.Y.	44.8	4.8			1.16	40.04	0.14	9.14		2879	311	H. W. Wurtz
" " Yonkers, N.Y.	52.76	4.11			2.05	35.88	0.77	4.43		2900	315	E. E. Moore
Terre Haute, Ind. (coal & water)	33.7	23.9		14.2	1.0	21.5		4.8		6775	732	W. A. Noyes
Wilkinson proc., Boston, Mass.	30.29	21.44		15.55	2.76	26.90		3.06		6372	682	Jenkins [& Schim'l
" Hoboken, N.J.	36.5	18.2	11.0	2.4	1.8	23.6		5.5	1.00	6243	668	Shepard, Bruckner
Hall process, Toledo, Ohio	9.5	5.0		1.5	8.5	14.0	1.5	60.0		1706	194	Smith
" "	9.9	7.3		5.5	7.7	14.0	1.3	54.3		3279	380	"

AIR AND WATER GAS.
EUROPE.

Name or Location.	Hydrogen.	Methane.	Ethylene.	Illuminants.	Carbonic Acid.	Carbonic Oxide.	Oxygen.	Nitrogen.	Sulphuretted Hydrogen.	Calories, per Cubic Meter.	B.T.U. per Cubic Foot.	Authority.
Dawson, Gas from Anth. coal	17.6				5.9	29.4		47.1		1363	146	Pfeiffer
" " " " "	17.0			2.0	6.0	23.0		52.0		1313	140	Schilling
" " " " "	18.4			0.6	7.2	26.8		47.0		1346	144	M. Witz
" " " Bit. coal	18.2			1.0	9.0	18.2		53.7		1015	108	"
" " " " "	21.9			0.7	4.4	45.9		50.7		1018	109	"
Anthracite	18.8		4.10		1.60	21.30		53.00		1548	164	"
Beure-Lecanchez	20.0			4.0	5.0	21.0		49.5		1262	135	"
Dowson										1443	156	Bueb Dessan
"	16.5			1.0	4.8	25.4		51.1		1287	138	Witz
Frankfort, Germany										2723	291	Bueb Dessan
Siemens, coal	16.85		2.4	2.05	4.55	22.75		53.80		1303	139	Ritchie
" St. Gobain	8.6				5.2	24.4		59.4		1283	1373	Kraus
" (after removing tar)	8.2				4.2	24.2		61.2		1357	145	Campbell, A.J.M.
" at Cambria, Pa.	5.3	3.2	0.2	2.2	4.0	22.8	5.6	63.4		1139	121	E., xxii, 375
" Midvale, Pa.	12		4.7		4.0	28		56.0		1212	129	Troilius
"	8.3		3.0		5.7	15.4		65.9		1167	125	"
"	6.0		2.4		1.5	23.6		65.9		1149	123	
"	9.8				5.9	17.7		64.2		1038	111	
Lecanchez, coke	10.83		1.10		3.57	21.76		61.36		1212	129	Witz
" dust, mixed	6.88		3.85		0.45	25.84		62.41		1362	145	"
" 12% volatile	20.00		3.50		5.00	21.00		49.50		1519	162	"
" dust, 14% volatile	12.27		4.62		2.96	22.91		55.62		1535	163	"
" 41% volatile	11.00		4.86		2.42	26.65		53.80		1542	163	"
Wilson, at Wrexham, England	0.90	0.91			6.91	29.58		61.7		1139	121	J. E. Stead

AIR AND WATER GAS.
EUROPE—Continued.

Name or Location.	Hydrogen.	Methane.	Ethylene.	Illuminants.	Carbonic Acid.	Carbonic Oxide.	Oxygen.	Nitrogen.	Sulphuretted Hydrogen.	Calories, per Cubic Meter.	B.T.U. per Cubic Foot.	Authority.
Wilson, at Wrexham, England.	1.11	1.43			8.29	26.33		62.84		1201	128	J. E. Stead
"	11.55	1.45			4.00	26.89		58.11		1136	121	"
Phillips, Midvale	8.7		1.4		3.9	27.30		67.4		940	100	Troilius
"	8.6		1.2		8.7	20.0		61.4		972	104	"
"	15.3		2.7		9.3	16.5		62.9		991	105	"
"	14.9		1.9		7.5	16.0		59.3		1115	119	"
"	28.7				8.0	15.5		61.6		926	98	"
Scotch	0.9		1.0		6.1	22.3		41.9		1642	175	J. L. Bell
Askam, England	0.14				21.7	29.24		48.2		946	101	"
Cleveland, "	0.06				13.47	33.80		52.59		859	92	"
Mond	27.7		1.8		14.32	27.03		58.84		669	72	
Kitson	1.5			0.4	17.0	11.0		42.5		1468	157	Ritchie
McKenzie	14.5	19.2		16.2	1.5	34.0		63.0		1495	160	
Gas from Wood	0.5	2.9	0.6		2.2	7.9	1.1	38.9		5280	565	W. A. Noyes
" Sawdust	0.8	2.3	0.3		11.5	28.4		56.1		1385	145	Anon
" Peat	0.9	2.7	0.4		15.1	26.1		55.4		1380	144	"
" Coal	0.65	2.5	0.45		12.1	27.2		56.7		1291	138	"
" Wood	0.7				7.35	28.0		61.0		1300	139	"
" Wood Charcoal	0.2				11.8	34.5		55.2		1156	124	Thwaite
" Peat	0.5				0.8	34.1		64.9		1120	119	"
" Coke	0.1				14.0	22.4		63.1		547	58	"
Furnace Gas from down comer	2.3	1.0			1.3	33.8		64.8		1131	121	Sweetser
"	1.2	1.2			12.4	25.8		58.5		2561	274	"
"					10.2	29.0		58.4		2505	268	

INDEX.

AGITATOR, BERTHELOT'S, 27
Aguitton's exp'ments on coal gas, 95
Air, analysis (table), 207
    necessary for combustion, 125; (table), 206
    necessary for combustion (table), 201, 202
    used in combustion, 139
Alexejew's calorimeter, 28
American Society of Mechanical Engineers, boiler-test report, 177
Analysis, Cinders, 115
  , Coal, 113
  , should show what, 114
  , Coke, 82
  , Gases, 133
  , Lignite, 78
  , Manchester gas, 93
  , Peat, 80
  , Proximate, 77
  , Waste gases (table), 134, 135
  , Wood, 84
Andrews' calorimeter, 47
Anemometer, Fan-wheel, 143
  , Fletcher's, 144
  , Volume of waste gases by, 143
Apparatus for steam-boiler testing should be correct, 183
  , Installation of, 13
  , Hirn's, 145
  , Orsat-Muencke, 134
Aqueous vapor, Heat of, 159
Ash, Analysis of, 115
  , Lignite, 78
  , Peat, 80
  , Treatment of, 189
Aspirator, Oil, 132
Atomic calorie, 2
Atwater's calorimeter, 71

BARRUS'S CALORIMETER, 38
Berthelot's agitator, 27
  bomb, 48

Bituminous schist, 79
Boghead coal, 79
Boiler-testing. See Steam-boiler Testing
Bomb. See Calorimeter
Briquettes, how made, 51
British thermal units, 2
  " " " to change to calories, 3
Brix's experiments with charcoal, 84
Bueb-Dessau's experiments on coal gas, 95
Bunsen's researches on flame, 168
Bunte's experiments on coal, 76
  gas-coke determinations, 9
  experiments on waste gases, 135
Burnat's smoke tests, 155

CALCULATION;
  Air necessary for combustion, 125
  Air supplied, 139
  Calories of the boiler test, 159
  Calories of carbon, 54
  Carpenter's calorimeter, 34
  Carbon, 54
  Coal, 66
  Coke, 68
  Colza oil, 64
  Favre and Silbermann's calorimeter, 26
  Flame temperature, 169
  Gases, 67, 94
  Heat units of boiler trial, 159
  Heat units by lead test, 10
  Heat units from chemical composition, 7
  Junker's calorimeter, 41
  Mahler's calorimeter, 61
  " " ; abridged, 70
  Regnault and Pfaundler's, 18
  Vapor of carbon, 173
  Volume of waste gases, 143
  Water value of calorimeters, 14, 63

250   INDEX.

Calculation ; Weight of waste gases, 141
Calories, atomic or molecular, 2
  Kilo-, 3
  Pound-, 2
  To change to B. T. U., 3. See Heat Units
Calorific power, 2
  Ratio of, to fixed carbon, 78
Calorimeter, Alexejew, 28
  Analytical, 74
  Andrews, 47
  Atwater, 71
  Barrus, 38
  Berthelot, 48
    corrections, 53
    examples, 54
    operation, 53
  Carpenter's, 31
    calculation, 34
  Constant pressure, 20
  Constant volume, 45
  Constant pressure and volume, ratio of, 45
  Correction for F. and S., 16
    Berthelot, 53
    cooling, 18, 60
    Junker's, 42
    Regnault and Pfaundler's, 18
  Cost of, 27
  Dulong, 20
  Evaluation in water. See Calorimeter, Water value
  Favre and Silbermann, 21
    Calculation, 26
    in complete combustion with, 23, 25
  Fischer, 29
  Hartley, 40
  Junker, 40
    calculation, 41
    errors, 42
  Kroeker, 73
  Mahler, 57
    and Berthelot compared, 70
    calculation, 61
    , abridged, 70
    enamel chips off, 58 (foot-note)
    examples, 64
    for gases, 62
    operation, 59
  Protection for, 13
  Rumford, 20
  Schwackhöfer, 35
    waste gases, 37
  Thompson, L., 43
  Thompson, W., 37

Calorimeter, Thomsen, 30
  Throttling, 117
  Walther-Hempel, 74
  Water value
    , Berthelot's calorimeter, 14
    by combustion, 14
    by mixing, 15
    Favre and Silbermann's calorimeter, 14
    Fischer's calorimeter, 30
    Lord and Haas' calorimeter, 14
    Mahler's calorimeter, 14, 63
    Witz, 47
Calorimeters, 12
Calorimetric endiometer, 47
Candle power and heat of combustion compared, 96
Cannel coal, 79
Carbon, calculation of calories, 54
  calories by various authors, 12
  in cinders, 115
  " smoke, 154
  "   "   ; analysis of, 154, 191
  oxygen necessary for, 125
  vapor, weight, and calories, 173
Carpenter's calorimeter, 31
Carbonic acid, Automatic determination of, 147, 148, 150
  in producer gases. See Gas Producer
  in waste gases, 81, 84, 91, 134, 137, 155
  , proper proportion in waste gases, 135
Carbonic oxide, Flame temperature of, 170
  in producer gas, 99
  in waste gases, 84, 91, 101, 134, 137 (table 135), 164
Cellulose, calories of, 85
Charbon roux, 83
Charcoal, peat, 80
  wood, 83
  ; Brix's tests, 84
  , half-burnt, 83
  ; Sauvage's tests, 83
  ; Scheurer-K.'s results, 84
  , Waste gases of, 84
Cinder, Analysis of, 115
Coal, Actual evaporation of, 76
  , Air necessary, 126
  , " supplied, 139
  Analysis, 113; (tables), 209–230
  "   should show, 114
  Bunte's experiments, 76
  Calories of, 66
  Difference in samples of, 113

Coal, Gruner's table, 77
   Heat of combustion (table), 198, 209
   Johnson's tests, 75
   Moisture in, 112, 114, 188
   Morin and Tresca's tests, 75
   Pure, 75
   Ratio of calories and fixed carbon, 77
   Ratio of hyd'gen and carbon, 78
   Sampling, 112
   Size for combustion, 24
   Uniformity in same bed, 112
   Weight of, 111
Coal gas. See Gas, Coal
Coke analyses (table), 209
   Calories of, 68
   Composition of, 82
   Heat of combustion (table), 230
   Kinds of, 81
   Use of, 82
Colza oil, Calories of, 64
Combustion. Air necessary, 125
   Air supplied, 139
   Heat of. See Heat of Combustion
   incomplete in F. and S. calorimeter, 23
Constant pressure, 20, 45
   " volume, 45
   " " relation of, to constant pressure, 45
Cooling, Newton's law, 60
   Regnault-Pfaundler's law, 18
Corrections for Berthelot calorimeter, 53
   Cooling, 18, 60
   Junker calorimeter, 42

DASYMETER, 146
Differential gauge, Segur's, 145
Dissociation, effect of, upon temperature, 168
Dulong's calorimeter, 20
Dulong's formula, 7
   , Agreement of, with test, 9
   , Mahler's limit to, 10 (foot-note)
   heat unit, 21

ECONOMETER, 148
Efficiency of steam-boilers, 191
Electric igniter, Heat of, 70
Evaluation in water. See Water Value
Evaporative effect of coal, 76
   , Factor for, 174
   power of fuel, 174

Evaporative power of charcoal, 84
   " " gas, 93
   " " lignite, 79
   " " peat, 80
   " " wood, 86
Evaporative power petroleum, 90
   of natural gas, 107
   unit, 180
Examples, Berthelot's cal'meter, 54
   Carpenter's calorimeter, 34
   Favre and S. " 26
   Mahler's " 64

FAN-WHEEL ANEMOMETER, 143
Favre and S.'s calorimeter, 21
Fischer's calorimeter, 29
Flame, 168
   Bunsen's researches, 168
   length, 169
   not due to incandescence, 168
   not due to solid particles, 168
   Propagation of, 168
   temperature, Calculation of, 169
   , Loss due to dissociation, 168
   acetylene, 170
   bor-methyl, 168
   carbon and carbonic oxide, 170
   hydrogen, 169
   marsh and olefiant gases, 171
   oils, 172
   petroleum, 172
   producer and other gases, 171
   solid fuels, 172
   table, 200
Fletcher's anemometer, 144
Flue-gas. See Waste Gases
Formula, Balling's, 8
   Burnat's, 143
   Dulong's, 7
   German Engineers', 8
   Hirn's, 146
   Jacobus's, 143
   Mahler's, 9
   Quality of steam, 119
   Regnault, for vaporization, 4
   Regnault and Pfaundler's, 18
   Schwackhöfer's, 8
   Superheated steam, 123
   Throttling calorimeter, 122
   Vaporization of water, 4
   Waste gases, weight, 141, 143
   Welter's, 10
Fuel, Air required for, 125; table, 206
   Air supplied to, 139
   Calorific power under steam-boiler, 109

Fuel, Evaporative power, 174
  Gaseous, 92
  Weight of, 111
Fuels, 1
  , Division of, 1
  Tables, 209

GAS, COAL
  Aguitton's experiments, 95
  Bueb-Dessau's experiments, 95
  Heat of combustion (table), 243
  Mahler's experiments, 96
  Variation in, 95
Gas-composimeter, 150
Gas, gasogene ; heat theory, 97
  Loss of calories, 98
  Value, 97
  Varieties, 98
Gas-holder, Oil, 132
Gas, Natural. See Natural Gas
Gas, Producer ; Heat theory of, 99
  Heat of combustion (table) 245, 246
  Mahler's experiments, 101
Gas sampler, A. S. M. E., 131
  Scheurer-Kestner's, 128
Gas, water. See Water Gas
Gaseous fuels, 92
  Heat of combustion of (tables), 245
Gases, Analysis, 133
  as fuel, 92
  Calculation of calories, 67
  Comparative value, 107
  Heat of combustion from analysis, 93
  Heat units, 164; table, 203
  " " example, 165
  Ignition point (table), 207
  Weight and volume (table), 200
  Specific heat (table), 204
Gases, waste. See Waste Gases
  Specific heat of (table), 205
Gottlieb's wood tests, 86
Gruener's coal table, 77

HARTLEY'S CALORIMETER, 40
Heat, balance in boiler trials, 193
  Loss of, in producer gas, 104
  of aqueous vapor, 159
    combination, 94
    combustible gases, 164
    combustion, 3
      and candle power, 96
      ; Calculated vs. det'mined, 9
      Cause of disagreement, 10
      Determination of, 3, 4
      From chem. composition, 7
      , Litharge or lead test, 10

Heat, Methods of determining, 7
  of carbon, 12, 54
  carbon vapor, 173
  coal, 66
  coke, 68
  colza oil, 64
  constant pressure, 20
  constant pressure and volume, 45
  fuels (tables), 209
  gas, 67
  gases, calculation, 68, 93
  gases, difference in, 94
  gases, modified by condensation, 94
  gases (table), 203, 241 *et seq.*
  hydrogen, 97
  marsh gas, 97
  natural gas, 106; table, 241
  oils (table), 238
  olefiant gas, 97
  petroleum, 90
  various subst. (table), 198
  electric igniter, 70
  hygroscopic water, 162
  sensible of the temperature, 160
  soot, 166
  vaporization of water, 4; table, 205
  water of combustion, 162
Specific ; gases (table), 204
  waste gases (table), 205
  water (table), 205
Heat units, Dulong's, 21
  from chemical composition, 7
  lead reduction test, 10
  Ratio of, to fixed carbon, 77
  of steam-boiler tests, Cal'tion, 159
  of steam-boiler tests Distribution, 167
Heat value, 2
  of fuels (tables), 209
Heating by charcoal, 84
  coke, 82
  gas, 92
  lignite, 78
  oil, 89, 90
  peat, 80
  wood, 84
Hirn's waste-gas apparatus, 145
  formula, 146
Horse-power, Commercial, 180
Hydrocarbons, Unconsumed, 25
Hydrogen, Calories of, 4
  in cinders, 115
  , Oxygen necessary for, 125

IGNITER, ELECTRIC
  Heat of, 70
Ignition point of gases (table), 207
Incandescence not flame, 168
Indiana natural gas analyses, 105
Installation of apparatus, 13

JACOBUS'S FORMULA, 143
Johnson's coal tests, 75
Junker's calorimeter, 40

KENT ON WASTE GASES, 141
Kent's ratio of hydrogen and carbon in coal, 78
  revision of Johnson's tests, 75
Kilo-calorie, 3
Kroeker calorimeter and correction for water, 73

LEAD OR LITHARGE TEST, 10
  is unreliable, 11
Lignite, 78
  , Heat of combustion (table), 231
Lord and Haas on Ohio and Pennsylvania coal, 9
Luminosity, 168
  depends on pressure, 169
  not due to solid particles, 168

MAHLER'S CALORIMETER, 57
  determinations of gas, 101
  experiments on coal gas, 96
  formula, 9
Manchester gas, Analysis of, 93
Mixed gas, 101
  , Calories of (table), 245
Moisture in coal, 112, 114
Moisture in steam, 119, 187
Molecular calorie, 2
Morin and Tresca on coal, 75
Morin and Tresca's wood tests, 86

NAPHTHALIN, CALORIES OF, 46
Natural gas and analysis of, 105
  Calories of, 106; (table), 241
  Value of, 106
  Variation in, 105
Nitrogen, ratio of, to oxygen (table), 207
Nixon's coal; calories of, determined, 66

OHIO NATURAL GAS, 105
Oil aspirator or gas-holder, 132
Oils, Heat of combustion (table), 238
Orsat-Muencke apparatus, 134

Oven cokes, Heat of combustion (table), 234
Oxygen, Compressed, is dry, 52
  in cylinders, 59
  necessary for combustion, 125
  "   "   "   (table), 201, 202
  , Ratio of, to nitrogen in air (table), 207
  required to form water with coal, 140; (table), 206
  To prepare, 24

PASTILLES, HOW MADE, 51
Peat, 80
  ; Calories of (table), 232
Petroleum, 88
  at Chicago, Canada, Moscow, 89
  , Calorific power of, 90
  heating tests, 90
  , Calories of (tables), 238
  , Steam used in atomizing, 91
Pittsburg natural gas, 105
Pneumatic pyrometer, 152
Pound-calorie, 2
Producer gas, 98. See Gas, Producer
Products of combustion of
  Alexejew's calorimeter, 28
  charcoal, 84
  Favre and Silbermann's calorimeter, 26
  oil, 91
  Schwackhöfer's calorimeter, 37.
    See Waste Gases.
Pyrometer, Pneumatic, 152

REGNAULT'S FORMULA, 4
Regnault and Pfaundler's law, 18
Ringelmann's smoke scale, 158
Ronchamp coal, Smoke of, 156
  "   "   Waste gases of, 134
Rothkohle, 83
Rumford's calorimeter, 20

SAMPLER, GAS, 128, 131
Sampling, Coal, 112
Sauvage's exp'ments on charcoal, 83
Scheurer-Kestner's experiments on charcoal, 84
  gas sampler, 128
  smoke analysis, 155
  and Meunier-Dollfus on coal, 75
Schist, Bituminous, 79
Schwackhöfer's calorimeter, 35
Segur's differential gauge, 145
Sensitiveness of thermometers, 6
Shale oil, 88

INDEX.

Smoke, Bunte's observations, 157
  Burnat's experiments, 155
  Carbon in, 154
  Ringelmann's scale, 158
  Scheurer-Kestner's analysis, 155
  Tatlock's tests, 155
Soda-lime for absorbing moisture, 23
Soot, Heat units of, 166
Specific heat. See Heat, Specific
  "   " of water not considered, 3
Steam, Moisture in, 117, 119, 187
  , Quality of, 119, 187
  , Superheated, 123
  , Temperature of, 116
  used in atomizing petroleum, 91
Steam-boilers, petroleum-fired, 89
  , Lignite-fired, 79
Steam-boiler testing
  apparatus to be correct, 183
  Ashes and residues, 189
  Analysis of cinders, 115
  "   " coal, 113
  "   " waste gases, 133, 190
  Boiler and chimney to be heated, 183
  Calculation of air necessary, 125
  "   " " supplied, 139
  "   " heat units, 159
  "   " waste gases, 136, 141, 146
  Carbon in smoke, 154
  Coal used, 182
  Corrections of apparatus, 183
  determine what, 109
  Distribution of calories, 167
  "   " heat, 109
  Duration of test, 115
  Early tests, 109
  Efficiency, 191
  Examination of boiler, etc., 182
  Heat balance. 192
  Heat tests and coal anal., 190
  Johnson's tests, 109
  Keeping records, 186
  Moisture in steam, 117
  Need of knowledge of calories in, 109
  Preliminaries of, 181
  Quality of steam, 119, 187
  Report of A. S. M. E. committee, 177
  Report of trial, 193
  Sampling the coal, 112
  Scheurer-Kestner's tests, 110
  Starting and stopping, 184

Steam-boiler testing, Temperature of steam, 116
  Temperature of waste gases, 151
  Volume of air necessary, 125
  "   " " supplied, 139
  "   " waste gases, 127
  Waste gas samples and analysis, 133, 190
  Water evaporated, 116
  Weight of fuel, 111
  "   " waste gases, 141
  What is necesary, 110
Sulphur, oxygen necessary for, 126

TABLE ; AIR COMPONENTS, 207
  Air for combustion, 201, 202
  " for perfect combustion, 206
  Ash analyses, 115
  Candle power and heat of combustion, 96
  Coal (Gruner's), 77
  Coke analyses, 82
  Distribution of calories, 167
  Flame temperatures, 200
  Fuels, 209
  Heat balance, 193
  Heat of combustion, 198
  "   " " of fuels, 209
  "   " " " gases, 202
  "   " " " lignites, 231
  "   " " " peat, 232
  "   " " " wood, 86, 233
  "   " vapor'n of water, 205
  Ignition point of gases, 207
  Natural gas, 105, 106, 241, 242
  Oxygen for combustion, 201, 202
  Oxygen to form water, 206
  Regnault and Pfaundler's law, 18
  Ronchamp coal waste gases, 134
  Smoke analyses, 157
  Specific heat of gases, 204
  "   " " waste gases, 205
  "   " " water, 205
  Thermometer reduction, 199
  Waste gas analyses, 134, 135
  Water value calculation, 15
  Weight and volume of gases, 200
  Wood, 86
Tatlock's smoke tests, 155
Temperature, Heat of sensible, 160
  " of waste gases, 151
Thermal units, 2
Thermometer, 4
  , Correction, mercury column, 6
  , Favre and Silbermann's, 6
  , Metastatic, 6
  , reduction table, 199

Thermometer, Sensibility of, 6
Thomsen's calorimeter, 30
Thompson's, L., calorimeter, 43
Thompson's, W., " 37
Throttling calorimeter, 117

UNIT OF EVAPORATION, 179
Units of heat, 3

VAPORIZATION OF WATER, 4
Vaporization of water (table), 205
Variation in coal gas, 95
" " natural gas, 105

WALTHER-HEMPEL
  Calorimeter, 74
Waste gas analysis, 190
Waste gases, Automatic apparatus for, 147
, Bunte's results, 135
from charcoal, 84
" petroleum, 91
" Ronchamp coal, 134
, Heat of, 160
, Hirn's apparatus, 145
" formula, 146
, Schwackhöfer's calorimeter, 37

Waste gases (table), 134, 135
, Temperature of, 151
Volume of, 127
Water evaporated, 116
, Heat of combination, 162
, Heat of vaporization of, 4: table, 205
, Hygroscopic, heat of, 162
in Lignite, 78
in peat, 80
, Kroeker's correction for, 73
, Specific heat (table), 205
, Specific heat of, not considered, 3
-value of cal'meters, 14, 15, 30, 63
Water gas, 101
, Heat of combustion of (table), 245 et seq.
Theory, 102
Loss of heat, 104
Weight of carbon vapor, 173
  fuel, 111
  waste gases, 141
Witz calorimeter, 47
Wood, Condition for burning, 87
  Gottlieb's tests, 86
  Calories (table), 86, 233
  Hydrate of carbon, 84
  Morin and Tresca's tests, 86
Wood charcoal. See Charcoal Wood.

www.ingramcontent.com/pod-product-compliance
Lightning Source LLC
Chambersburg PA
CBHW032000230426
43672CB00010B/2216